Analog Circuits and Signal P

Series Editors:
Mohammed Ismail
Dublin, USA

Mohamad Sawan
Montreal, Canada

For further volumes:
http://www.springer.com/series/7381

Khaled Salah • Yehea Ismail • Alaa El-Rouby

Arbitrary Modeling of TSVs for 3D Integrated Circuits

Springer

Khaled Salah
Mentor Graphics
Heliopolis
Egypt

Alaa El-Rouby
Mentor Graphics
Heliopolis
Egypt

Yehea Ismail
The American University (New Cairo)
and Zewail City of Science and Technology
(6th of October City)
New Cairo
Egypt

ISBN 978-3-319-37497-0 ISBN 978-3-319-07611-9 (eBook)
DOI 10.1007/978-3-319-07611-9
Springer Cham Heidelberg New York Dordrecht London

Softcover reprint of the hardcover 1st edition 2015

Printed on acid-free paper

Springer is part of Springer Science+Business Media (www.springer.com)

Preface

This book presents a wide-band and technology independent SPICE-compatible *RLC* model for TSVs in 3D ICs. This proposed model accounts for a variety of effects, including skin effect, depletion capacitance, and nearby contact effects. The TSV is modeled like a MOS structure where it is assumed that a full depletion region exists around the TSV. A lumped parameter model is then proposed to model the TSV. The equivalent circuit model includes a single TSV model and coupling model between TSVs. The dimensional analysis method is applied to obtain closed form solutions for the resistance, capacitance, and inductance of the TSV lumped model. The accuracy of the expressions is then verified with the electromagnetic field solver under typical high-density TSV dimensions, and it shows a significant accuracy up to 100 GHz.

Although there are several works in the literature that provide an *RLC* model for a TSV and closed form expressions with different levels of accuracy, this book discusses models that exhibit several additional enhancements as compared to existing literature. The models in this book include: (1) MOS depletion R and C effect. (2) Body contact effect. (3) Model linearization, i.e., single nonlinear or frequency-dependent element can be approximated by multiple linear, and frequency-independent elements. (4) Simulation comparisons (e.g. with full-wave, quasi-static, and device simulation). This book can be very useful in the fast SPICE-compatible parasitic extraction of TSVs for 3D IC design.

Moreover, a proposed architecture based on TSV technology for a spiral inductor is demonstrated and characterized. Also, in this book, and for the first time, a novel inductive coupling interface that uses the magnetic near field induced by TSV-based spiral inductor is demonstrated. The feasibility of using TSV for wireless near field communication is shown. A TSV-based near-field inductive-coupling system offers a high quality factor and a good coupling coefficient. Therefore, the proposed communication system appears to be a promising technology for wireless communication. Moreover, another application for the TSV, which is bandpass filter, is discussed in this book.

This book presents the effects of substrate doping density on the electrical performance of TSV. Moreover, parasitic coupling capacitive between through-silicon vias and surrounding wires is analyzed and studied. Noise coupling between TSVs and CMOS is also investigated.

This book also presents several new directions for TSV fabrication and use in design. It presents performance comparison between Air-Gap Based Coaxial TSV and conventional circular TSV. Air-gap TSVs reduce the energy loss compared to the conventional circular TSV or conventional coaxial TSV. Moreover, SW-CNT bundles as a prospective filler material for TSV are investigated compared to conventional filler materials like Cu, W, and poly-silicon. CNT-filled TSVs have superior performance compared to Cu-filled TSVs, resulting in reduction in transmission loss at high frequency of operation. Moreover, TSV-based ADIABATIC logic based on the adiabatic switching principle is presented and analyzed.

Contents

Chapter 1
Introduction: Work Around Moore's Law

Interconnect dimensions and complementary metal–oxide–semiconductor (CMOS) transistor feature sizes approach their physical limits. Therefore, scaling will no longer be the sole contributor to performance improvement. In addition to trying to improve the performance of traditional CMOS circuits, integration of multiple technologies and different components in a high-performance heterogeneous system is a major trend. This chapter briefly surveys key technology level trends, classified as "More Moore" such as: new architectures (silicon on insulator, SOI; FinFET; Twin-Well) and new materials (High-*K*, metal gate, strained Si), "More than Moore" such as: new interconnect schemes (three-dimensional, 3D; network on chip, NoC; optical; wireless), and "Beyond CMOS" such as: new devices (molecular computer, biological computer, quantum computer).

1.1 Scaling Limitations of Conventional Integration Technology: Work Around Moore's Law

By 2020, the minimum physical gate length of transistors will be close to 7 nm (Fig. 1.1), which is considered by most researchers to be the physical limit of silicon as that limit is the size of the atom and molecule. Clearly, devices cannot be fabricated smaller than the dimension of a single molecule. Also, lithography technology seems unfit for precise manufacturing beyond 7 nm, and if silicon dioxide insulators are reduced to just a few atomic layers, electrons can tunnel directly through the gate. The interconnect congestion bottleneck is also a limiting factor (Fig. 1.2), where interconnect delay dominates with scaling (Table 1.1). These limitations of silicon-based integrated circuits (ICs) are summarized in Table 1.2. These limitations are now causing the industry to identify at least three main research domains (Fig. 1.3), namely: (1) "More Moore," (2) "More than Moore," and (3) "Beyond CMOS" (CMOS, complementary metal–oxide–semiconductor) [1].

The "More Moore" domain traditionally deals with technologies related to further scaling silicon-based CMOS. The "More than Moore" domain encompasses the engineering of complex systems that can combine, by heterogeneous integra-

K. Salah et al., *Arbitrary Modeling of TSVs for 3D Integrated Circuits*,
Analog Circuits and Signal Processing, DOI 10.1007/978-3-319-07611-9_1,

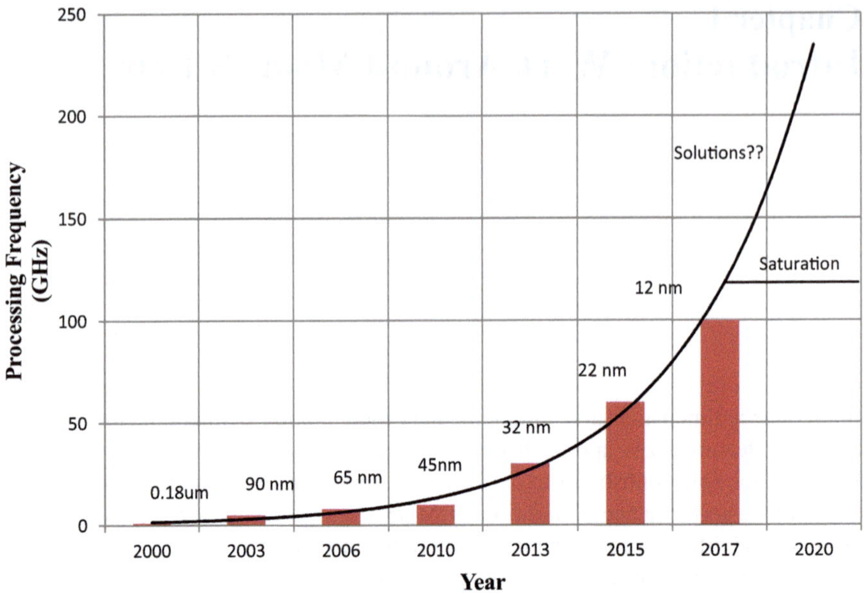

Fig. 1.1 Technology trends show saturation in transistor scaling at 12 nm and the need for working around Moore's law [1]

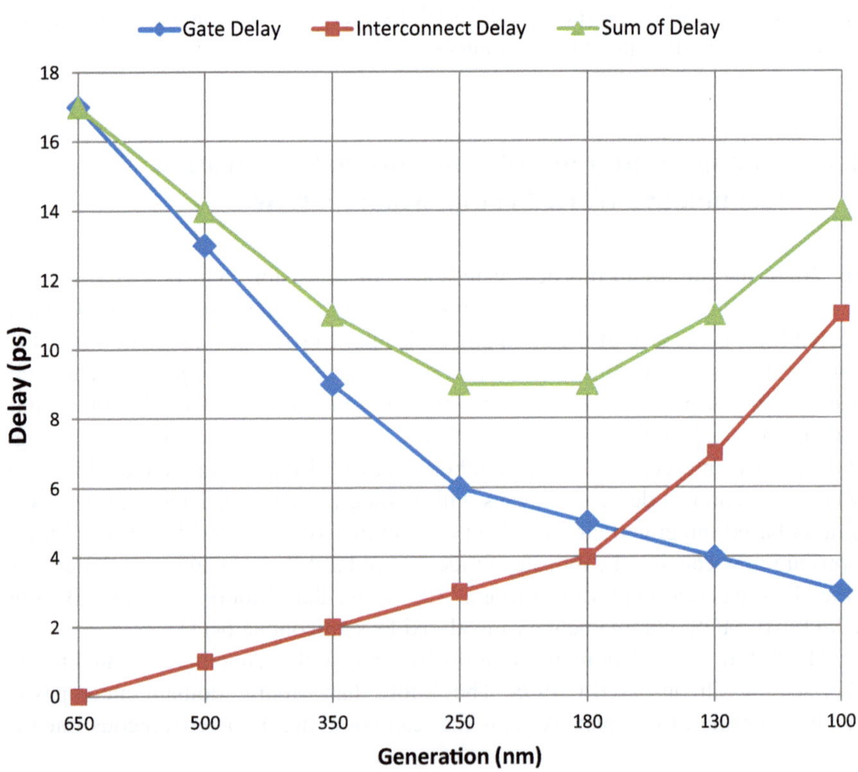

Fig. 1.2 Gate delay, interconnect delay, and sum of delay at different technology nodes, which shows the fast increase in delay time caused by the interconnect [1]

Table 1.1 Interconnects dominate with scaling [1]

	90 nm	45 nm	22 nm	12 nm
Transistor delay (ps)	1.6	0.8	0.4	0.2
Delay of 1 mm interconnect (ps)	5×10^2	2×10^2	1×10^4	6×10^4
Ratio	3×10^2	3×10^3	4×10^4	3×10^5

Table 1.2 Physical limitations of silicon-based ICs

Manufacturing limitations	
Lithography	Lithography technology seems unfit for precise manufacturing beyond 7 nm
Transistor dimensions	Transistor dimensions are approaching a hard limit that cannot be overcome That limit is the size of the atom and molecule. Clearly, devices cannot be fabricated smaller than the dimension of a single molecule [5]
Material limitations	
SiO_2	If silicon dioxide insulators are reduced to just a few atomic layers, electrons can tunnel directly through the gate
Design limitations	
Interconnects bottleneck	Interconnect delay dominates over gate delay

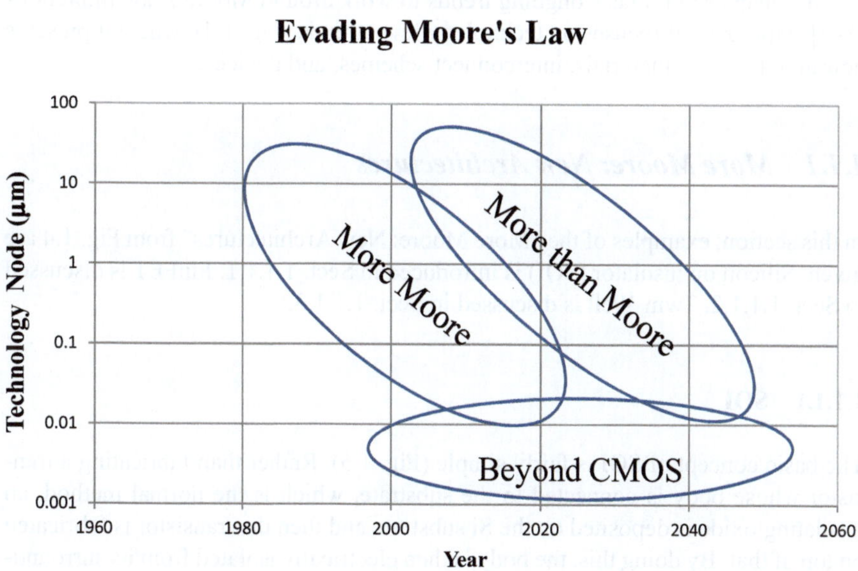

Fig. 1.3 Evading Moore's law: *1* "More Moore," *2* "More than Moore," and *3* "Beyond CMOS" [4]

tion techniques (in SoC or SIP), various technologies (not exclusively electronic) in order to meet certain needs and challenging specifications of advanced applications. The "Beyond CMOS" domain deals with new technologies and device principles (i.e., from charge-based to non-charge-based devices, from semiconductor to molecular technologies).

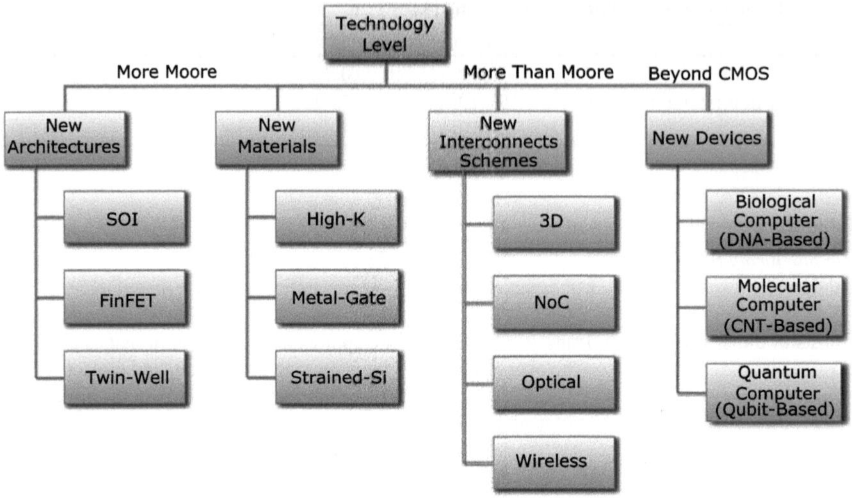

Fig. 1.4 Summary of solutions on the technology level to work around Moore's law, where it presents new architectures, materials, interconnect schemes, and devices

This chapter introduces ongoing trends to work around Moore's law limitations [1–4]. Moreover, it focuses on technology level trends (Fig. 1.4), where it presents new architectures, materials, interconnect schemes, and devices.

1.1.1 More Moore: New Architectures

In this section, examples of the "More Moore: New Architectures" from Fig. 1.4 are given. Silicon on insulator (SOI) is introduced in Sect. 1.1.1.1. FinFET is discussed in Sect. 1.1.1.2. Twin-Well is discussed in Sect. 1.1.1.3.

1.1.1.1 SOI

The basic concept of SOI is fairly simple (Fig. 1.5). Rather than fabricating a transistor whose body is connected to the substrate, which is the normal method, an insulating oxide is deposited on the Si substrate and then the transistor is fabricated on top of that. By doing this, the body is then electrically isolated from its surroundings. This means that the bulk-to-source voltage is now floating which lowers the threshold voltage and capacitance, resulting in a performance increase [5].

1.1.1.2 FinFET

FinFETs have their technology roots in the 1990s, when DARPA funded research into possible successors to the planar transistor. FinFET, which is a double-gate field

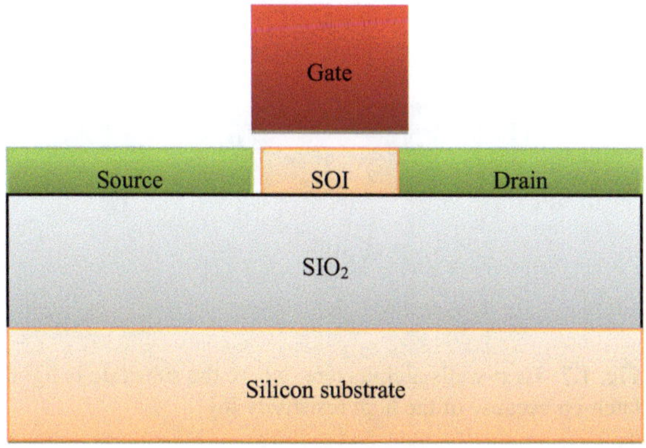

Fig. 1.5 SOI CMOS architecture, where rather than fabricating a transistor whose body is connected to the substrate, which is the normal method, an insulating oxide is deposited on the Si substrate and then the transistor is fabricated on top of that [5]

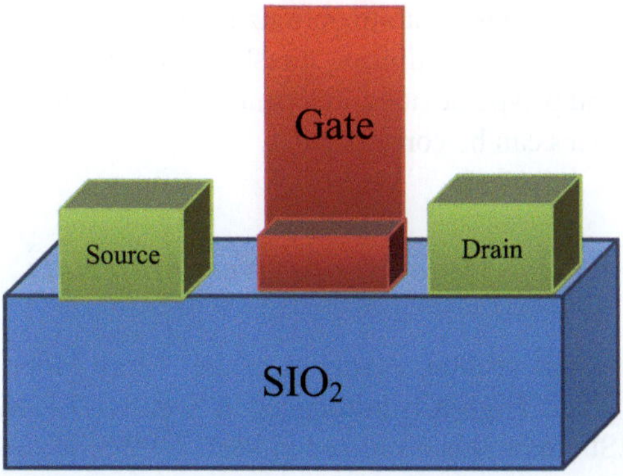

Fig. 1.6 FinFET architecture, where it has two gates that can be controlled independently, usually, the second gate of FinFETs is used to dynamically control the threshold voltage of the first gate in order to improve circuit performance and reduce leakage power [6]

effect transistor (DGFET), is more versatile than traditional single-gate field effect transistors because it has two gates that can be controlled independently (Fig. 1.6). Usually, the second gate of FinFETs is used to dynamically control the threshold voltage of the first gate in order to improve circuit performance and reduce leakage power. Also, a FinFET can be considered a three-dimensional (3D) version of metal–oxide–semiconductor field-effect transistor (MOSFET) that rises above the planar substrate, giving them more volume than a planar gate for the same planar area. Given the excellent control of the conducting channel by the gate, which "wraps" around the channel, very little current is allowed to leak through the body when the device is in the off state. This allows the use of lower threshold voltages, which results in optimal switching speeds and power [6].

1.1.1.3 Twin-Well

For high-performance chips, a low-doped substrate is used, on which two wells are constructed at optimum doping levels. Since the substrate is lightly doped, there

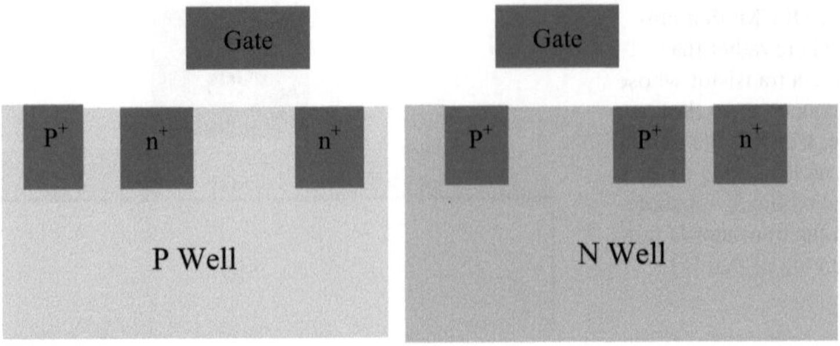

Fig. 1.7 Twin-wells architecture, where the substrate is lightly doped, so there is less chance for latch-up because of the high resistivity [6]

is less chance for latch-up because of the high resistivity (Fig. 1.7) [6]. Moreover, it provides separate optimization of the n-type and p-type transistors and makes it possible to optimize "threshold voltage," "body effect," and the "gain" of n-type and p-type devices, independently. Therefore, balanced performance of n and p devices can be constructed.

1.1.2 More Moore: New Materials

In this section, examples of the "More Moore: New Materials" from Fig. 1.4 are given. High-*K* is introduced in Sect. 1.1.2.1. Meta gate is discussed in Sect. 1.1.2.2. Strained Si is discussed in Sect. 1.1.2.3.

1.1.2.1 High-K Dielectric

The term high-*K* dielectric refers to a material with a high dielectric constant *K* (as compared to silicon dioxide). High-*K* dielectrics are used in semiconductor manufacturing processes where they are usually used to replace a silicon dioxide gate dielectric or another dielectric layer of a device. High-*K* dielectrics are designed to address one particular aspect of off-state power consumptions: gate-tunneling currents, as it reduces gate leakage [4].

1.1.2.2 Metal Gate

The doped polycrystalline silicon used for gates has a very thin depletion layer, approximately 1 nm thick, which causes scaling problems for small devices. Others metals are being investigated for replacing the silicon gates, including tungsten and molybdenum to eliminate the polysilicon depletion limitation and also for better thermal stability [4].

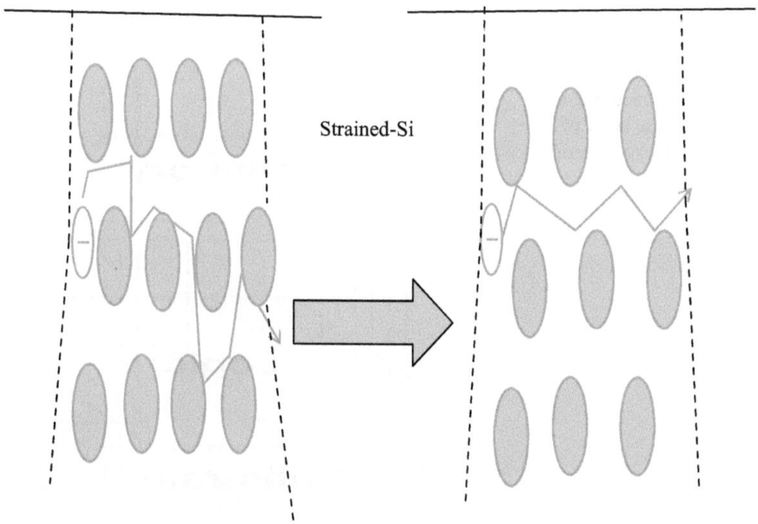

Fig. 1.8 Strained Si is the process of introducing physical strain on the silicon lattice to help improve electron and hole mobility [5]

1.1.2.3 Strained Si

If we cannot make MOSFETs yet smaller, instead move the electrons faster. Strained Si is the process of introducing physical strain on the silicon lattice to help improve electron and hole mobility (Fig. 1.8). This allows the holes and electrons to flow more freely, reducing device resistance and other properties affected by electron/hole mobility [5].

1.1.3 More than Moore (MTM): New Interconnects Schemes

In this section, concepts of "More Moore: New Interconnect Schemes" from Fig. 1.4 are listed. 3D interconnect is introduced in Sect. 1.1.3.1. Network on chip (NoC) is discussed in Sect. 1.1.3.2. Optical interconnect is discussed in Sect. 1.1.3.3. Wireless interconnects are discussed in Sect. 1.1.3.4.

1.1.3.1 3D

Increasing drive for the integration of disparate signals (digital; analog; radiofrequency, RF) and technologies (SOI, SiGe heterojunction bipolar transistors (HBTs),

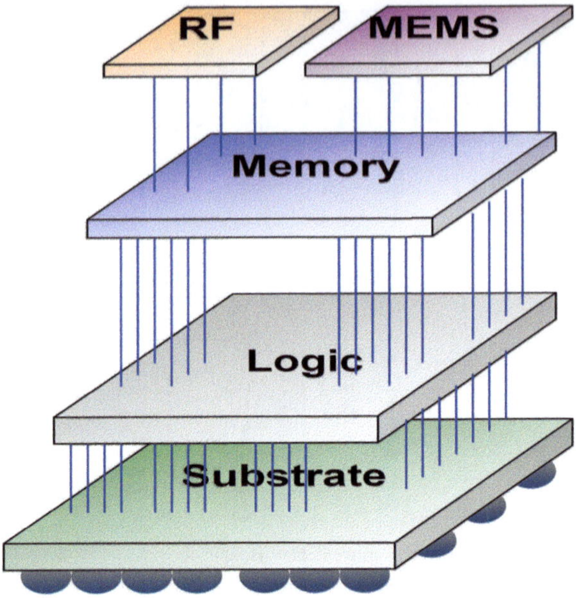

Fig. 1.9 3D technology, where an entire (3D) chip is divided into a number of different blocks, and each one is placed on a separate layer of silicon that are stacked on top of each other [7]

GaAs, strained silicon, and so on) results in introducing various design concepts, for which existing planar technologies may not be suitable; hence, 3D-ICs are introduced (Fig. 1.9).

3D integration technology provides increased performance in many design criteria as compared to the current 2D approaches. 3D-ICs, which contain multiple layers of active devices, extensively utilize the vertical dimension to connect components and are expected to address interconnect delay-related problems in planar (3D) technologies, by the use of short wires in 3D designs. These shorter wires will decrease the average load capacitance and resistance and decrease the number of repeaters which are needed to regenerate signals on long wires; 3D technology also enables the integration of heterogeneous technologies.

In the 3D design, an entire (3D) chip is divided into a number of different blocks, and each one is placed on a separate layer of silicon that are stacked on top of each other. This may be exploited to build a system on chip (SoC) by placing different circuits with different performance requirements in different layers [7].

3D integration provides some unique benefits for processor design, such as packing density, interconnect bandwidth and latency, modularity, and heterogeneity. The packaging density improvement provided by 3D can be used to continue improvements in processing and storage capacity as well as enabling a gradual shift towards integrating the full system in one stack.

3D enables modular design of a variety of systems from a shared set of subcomponents through functional separation of the device layers. As a result, different layers can be independently manufactured in the most cost-effective ways, which can be stacked to compose a wide range of customized systems. Optimizing layer

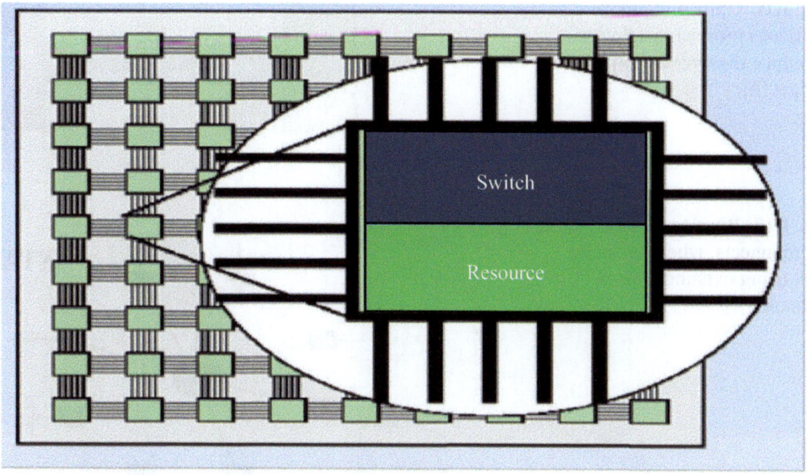

Fig. 1.10 NoC architecture. Instead of having a shared bus, a message passing approach is used [7]

interfaces and infrastructure components, such as power delivery and clocking, can further enhance the inherent modularity advantages.

The system-level benefits of 3D will be determined, to a significant degree, by the effectiveness of novel design methodologies that explore the new design space introduced by the vertical dimension. Design flow optimization is essential in achieving the highest performance gains as well as tackling the more prominent interdependencies among performance, power dissipation, temperatures, interconnectivity, and reliability in 3D.

1.1.3.2 NoC

The NoC is a promising solution to simplify and optimize SoC design. NoCs have been introduced as a shared communication medium that is highly scalable and can offer enough bandwidth to replace many traditional bus-based and point-to-point links [8]. NoC is a communication infrastructure for complex SoC systems with many IPs like a multiprocessor system. Instead of having a shared bus, a message passing approach is used. In analogy to a computer network, each IP acts like a processor, sending and receiving packets (flits) to and from the network. The basic element of a NoC is the switch, which connected to the IP with a network interface (NI), that is in charge of routing flits from the input ports to the output ports, from the source IP to the destination IP (Fig. 1.10). When a flit reaches its destination, it will have crossed several switches, i.e., hops. This kind of communication infrastructure is scalable, making it possible to have a number of parallel communications between different cores.

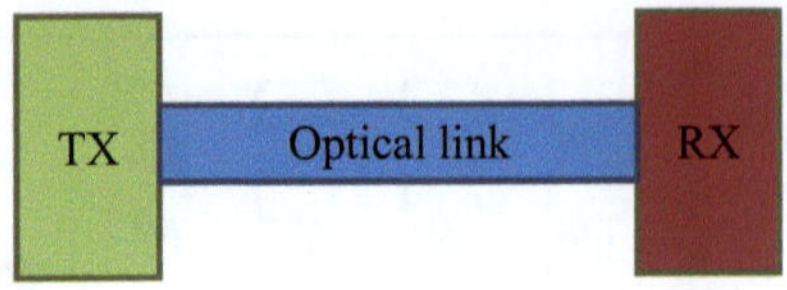

Fig. 1.11 Optical interconnects as a promising solution to reduce the overall power budget [6]

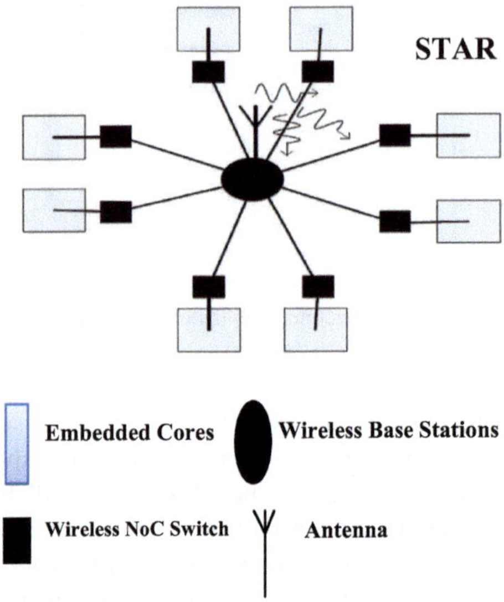

Fig. 1.12 Wireless NoC interconnects, where it can be built using existing CMOS technology [10]

1.1.3.3 Optical Interconnects

Thanks to the unique properties of optical communication, such as bit-rate transparency and low-loss optical waveguides, photonic NoCs have been introduced as a promising solution to reduce the overall power budget (Fig. 1.11) [6].

1.1.3.4 Wireless Interconnects

Unlike 3D and photonic NoCs, NoC with RF interconnects can be built using existing CMOS technology (Fig. 1.12) [10]. Antennas are just on-chip interconnect with transceivers also on-chip enabling full core-to-core communication as well as wireless clocking. Table 1.3 summarizes the main differences between different interconnect technologies.

1.1.4 Beyond CMOS: New Devices

In this section, examples of "Beyond CMOS: New Device Schemes" are discussed. CNT is introduced in Sect. 1.1.4.1. Biological computers are discussed in Sect. 1.1.4.2. Quantum computers are discussed in Sect. 1.1.4.3.

Table 1.3 A comparison between new interconnect schemes

	3D & NoC	Optical	Wireless
Design requirements	Multiple layers with active devices	Silicon photonic components	On-chip metal or CNT-based antennas
Performance gain	Higher connectivity Shorter average path length	High-speed optical devices and links Negligible power dissipation	Direct point-to-point wireless links between smaller subnets Multi-hop paths replaced by single hop links
Reliability	Vertical via failure	Temperature sensitivity of photonic components	Noisy wireless channel
Challenges	Heat dissipation due to higher power density Yield	Integration of on-chip photonic components	Low-power mm-wave transceivers

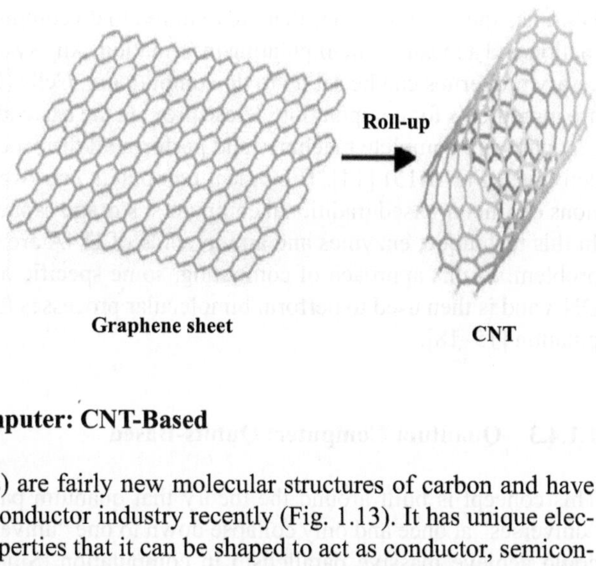

Fig. 1.13 SW-CNT structure, where CNT is cylinders of graphite, same as the material in a pencil. CNT can be metallic or semiconducting; it all depends on the angle of rolling and diameter of the nanotube [9]

1.1.4.1 Molecular Computer: CNT-Based

Carbon nanotubes (CNTs) are fairly new molecular structures of carbon and have been applied to the semiconductor industry recently (Fig. 1.13). It has unique electrical and mechanical properties that it can be shaped to act as conductor, semiconductor, or insulator. CNT is able to reduce semiconductor manufacturing processing steps that incorporate copper, such as the forming of insulator to prevent copper from diffusion through the gate. CNTs are cylindrical structures based on the hexagonal lattice of carbon atoms that covalently bond together and forms crystalline graphite. CNTs can be looked at as single molecules, due to their small size (~ 1 nm in diameter and ~ 1-μm length) [11–13].

CNTs are cylinders of graphite, same as the material in a pencil. CNT can be metallic or semiconducting; it all depends on the angle of rolling and diameter of the nanotube. CNT-based through-silicon via (TSV) has excellent current-carrying capability higher than copper, as it does not exhibit skin effect at high frequency (Cu, 10^6 A/cm^2; CNT, 10^9 A/cm^2). Moreover, it has better thermal stability than

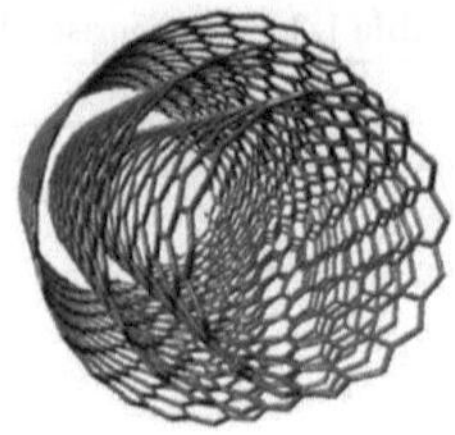

Fig. 1.14 MW-CNT is an assembly of concentric single-walled carbon nanotubes with different diameters [9]

copper as it works fine at much higher temperatures, thus eliminating the problem of electromigration with Cu interconnects.

There are two major types of CNTs: single-walled carbon nanotubes (SW-CNTs) and multi-walled carbon nanotubes (MW-CNTs) [15]. MW-CNTs can be described as an assembly of concentric SW-CNTs with different diameters (Fig. 1.14).

1.1.4.2 Biological Computer: DNA-Based

DNA has the two properties that are required to do computing: a way to store information and a means of manipulating information. Any system having these two necessary properties can be set up to do computation. Cells (living parts) of organisms are ingredients for computation. These provide the basic idea of computing, as these tiny parts are complete machines and perform all the processing for the organisms' activities (Fig. 1.15) [14]. Biological computers can overcome two major limitations of silicon-based traditional computers: storage capacity and processing speed. In this technique, enzymes and amino acids of DNA are used to solve a particular problem. In this approach of computing, some specific information is encoded on DNA and is then used to perform bimolecular processes to attain the intended computation [15–18].

1.1.4.3 Quantum Computer: Qubits-Based

This concept is built around the theory that quantum particles can exist in many "universes" at once and only collapse down to one "universe" when observed. This could achieve massive parallelism in computation. Silicon-based computers use transistors to build logic gates but quantum computers use special quantum bits called qubits. Moreover, a group of qubits is called qubit register. Traditional computer bits have only two states either "0" or "1" but on the other hand a quantum computer can carry multiple values at the same time. Qubit can be either "0" or "1" and sometimes both "0" and "1" at the same time which is known as superposition of states. Quantum computation can be used to simplify a variety of problems, such as search algorithms and cryptography number theory. Nevertheless, quantum computers cannot solve many other problems such as sorting data [15–19]. An example of an implemented quantum computer is shown in Fig. 1.16.

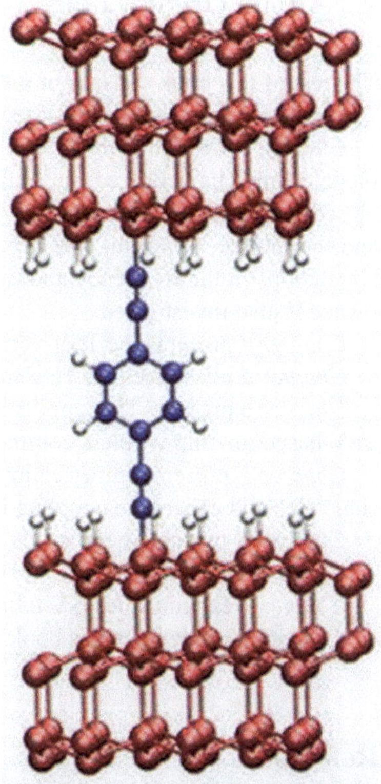

Fig. 1.15 DNA-based biological computer: the research by Bernholc and Acutes has shown that a benzene molecule, in *blue*, can act as a very fast switch in a transistor when connected to silicon leads, in *red* [13]

Fig. 1.16 A 16-qubit quantum computer. Silicon-based computers use transistors to build logic gates but quantum computers use special quantum bits called qubits [15]

1.2 Book Organization

The rest of this book consists of the following chapters.

Chapter 2 discusses the TSV technology, structure, challenges, and fabrication.

Chapter 3 describes the TSV modeling methodology. The TSV is physically modeled and characterized, and different multi-TSV structures are investigated and closed-form formula is introduced. Also, the TSV model linearization is introduced, and the models for multi-TSV arrangements are validated versus electromagnetic simulations in the frequency and time domains. The TSV impact on circuit performance is also investigated.

Chapter 4 presents the TSV verification methodology.

Chapter 5 introduces the TSV macro-modeling framework.

Chapter 6 presents TSV applications such as TSV-based on-chip spiral inductor, TSV-based on-chip wireless communications, and TSV-based bandpass filter.

Chapter 7 presents imperfection in TSV modeling which includes nonuniform substrates and capacitive coupling between metal lines in 3D-ICs.

Chapter 8 presents new trends in TSV which includes air-gap TSV, adiabatic TSV, and CNT-TSV.

Chapter 9 explains the TSV fabrication process.

Chapter 10 concludes this book.

References

1. The International Technology Roadmap for Semiconductors (ITRS), 2012 [Online]. http://www.itrs.net. Accessed 2013
2. K. Roy, B. Jung, A. Raghunathan, Integrated systems in the More-than-Moore era: designing low-cost energy-efficient systems using heterogeneous components, in *23rd International Conference on VLSI Design*, 2010
3. Z. Lu, A. Jantsch, Trends of terascale computing chips in the next ten years, ASICON, 2009
4. A.M. Ionescu, Nanoelectronics roadmap: evading Moore's law, EWME, 2009
5. D. Rairigh, Limits of CMOS technology scaling and technologies Beyond-CMOS, IEEE, 2005
6. M.C. Wang, Independent-gate FinFET circuit design methodology. IAENG Int. J. Comp. Sci. **37**, 1 (2010)
7. R. Weerasekera, M. Grange, D. Pamunuwa, H. Tenhunen, L.-R. Zheng, Compact modelling of through-silicon vias (TSVs) in three dimensional (3-D) integrated circuit, in *Proceeding of IEEE International Conference on 3D System Integration (3D IC)*, San Francisco, USA, 2009
8. L.P. Carloni, Y. Xie, Networks-on-chip in emerging interconnect paradigms: advantages and challenges, IEEE, 2009
9. D. Goswami, Optical computing. Resonance **8**(6), 56–71 (2003)
10. K. Bergman, L.P. Carloni, J.A. Kash, Y. Vlasov, On-chip photonic communication for high-performance multi-core processors, IEEE, 2009
11. A. Gupta, B.C. Kim, S. Kannan, S.S. Evana, L. Li, Analysis of CNT based 3D TSV for emerging RF applications, in *Electronic Components and Technology Conference*, 2011
12. B. Kim, S. Kannan, A. Gupta, S. Noh, L. Li, Characterization of high performance CNT-based TSV for radar applications, in *IEEE 13th Electronics Packaging Technology Conference*, 2011

13. S. Kannan, B. Kim, A. Gupta, S. Noh, L. Li, Characterization of high performance CNT-based TSV for high-frequency RF applications. Adv. Mater. Res. **1**(1), 37–49 (2012)
14. http://www.physorg.com/news90607516.html. Accessed 2013
15. S. Hassan, M. Asghar, Limitation of silicon based computation and future prospects, in *Second International Conference on Communication Software and Networks*, 2010
16. J.F. Podevin, T. Munakata, Beyond silicon new computing paradigms. Commun. ACM **50**(9), 30–34 (2007)
17. S. Khullar, V. Chopra, M.S. Kahlon, DNA computing: migrating from silicon chip to test tubes, in *Proceeding of National Conference on Challenges & Opportunity in Information Technology, COIT 2007*, March 23 (RIMT-IET, Mandi Gobindgarh, 2007), pp. 72–75
18. L.M. Adleman, Computing with DNA. Sci. Am. **279**(2), 54–61 (1998)
19. Z. Ramjan, Quantum computing, CS 664 (Spring 2005), http://zack.ramjanfamily.com/cs664/Quantum_Paper.pdf. Accessed Oct 2009

Chapter 2
3D/TSV-Enabling Technologies

Three-dimensional (3D) integration is a promising alternative option to traditional two-dimensional (2D) planar chips. The 3D integration is mainly concerned with the communication infrastructure between different stacked dies of future multi-core system-on-chip (SoC) and network-on-chip (NoC). Among several 3D integration technologies, the through silicon via (TSV) approach is the most promising one and therefore is the focus of the majority of 3D integration R&D activities. However, there are challenges that should be overcome before the production of TSV-based 3D integrated circuits (ICs) becomes possible, e.g., electrical modeling challenges, thermal and power challenges, technological challenges, design methodology challenges, and computer-aided design (CAD) tool development challenges. The manufacturability of TSV-based 3D-ICs is an important issue for realizing real 3D-ICs designs.

2.1 3D/TSV Technology

The small-scale production of integrated circuits (ICs) started in the early 1960s. The modern-day ICs may have billions of transistors. The functionality scales by reducing the size of the transistors and also by increasing the transistor count per unit area of the processor die. More transistors are added to processors with each technology node. System-on-chip (SoC) was introduced in order to integrate all the components on a single substrate. SoC has incredible functionality, having processor, digital logic, memory, embedded intelligence, and analog components on a single die. However, interconnect length is a critical issue for SoC due to increase in the chip dimensions. Mixed signal integration is another issue for present-day SoC as the noisy digital part cannot be placed in close proximity to the sensitive analog part. In addition to this, same technology node has to be followed for different components of SoC which may place stringent requirements on analog and radio frequency (RF) components. Three-dimensional (3D) integration with through silicon vias (TSVs) is, therefore, an obvious choice for mixed technology applications compared to SoC. In other words, increasing drive for the integration of different

K. Salah et al., *Arbitrary Modeling of TSVs for 3D Integrated Circuits*,
Analog Circuits and Signal Processing, DOI 10.1007/978-3-319-07611-9_2,

Table 2.1 A comparison between alternative design implementations

Property	Single chips	SoC	3D
Modular flexibility	High	Low	Medium
System performance	Low	Medium	High
Physical dimension	Large	Medium	Small
Complexity of fabrication	Low	Medium	Medium
Cost	Low	Medium	High
CAD tools and design methodologies	Available	Available	Not deployed yet

signals (analog, digital, RF) and technologies (SOI, HBTs, GaAs, strained silicon, and so on) is introducing various design concepts, for which existing planar (3D) technologies may not be suitable, so 3D-ICs are introduced.

The purpose of 3D integration is either to partition a single chip into multiple strata to reduce global interconnect length or stacking of chips together using TSVs. Increasing the number of strata from one to four reduces the length of the longest interconnect by 50% with 75% improvement in latency and 50% improvement in interconnect energy dissipation. Using 3D integration, the clock frequency can be increased by 3.9× and area and power can be reduced by 84%. Energy dissipation of long on-chip wires can be reduced by 54% by using 3D interconnects in a 45-nm technology node. 3D integration provides the potential for a tremendously increased level of integration per unit footprint compared to its two-dimensional (2D) counterpart.

The migration from 2D to 3D-ICs brings higher bandwidth, reduced power consumption, and higher integration density. The improved performance, reduced timing, and small form factors are also key drivers for 3D integration and TSV interconnect technology.

A qualitative comparison regarding the benefits introduced by the 3D integration process, compared to existing system design approaches, is summarized in Table 2.1 and illustrated in Fig. 2.1.

Another classification to technology is divided into SiP, SoP, SoC, and 3D, where system-in-package (SiP) can hire different chips/dies from different processes in one package. This can be quite good, for example, where you have to get complementary metal oxide semiconductor (CMOS)-based parts and high-frequency (HF) components in one package (wireless apps). The latter is quite difficult to do in CMOS process. We use SiP for chips integrating different kinds of technologies like RF, SOI, etc. making the coexistence of these different technologies on the same chip possible. SiP is based on bonding wires (Fig. 2.2). System-on-package (SoP) is a "technical" term for printed circuit board (PCB) with bare dies in a single package (Fig. 2.3). The 3D approach has a considerable edge in feature size compared to SoP as shown in Table 2.2.

3D integration technology provides increased performance in many design criteria as compared to the current 2D approaches, since 3D-ICs, which contain multiple layers of active devices, extensively utilize the vertical dimension to connect components and are expected to address interconnect delay-related prob-

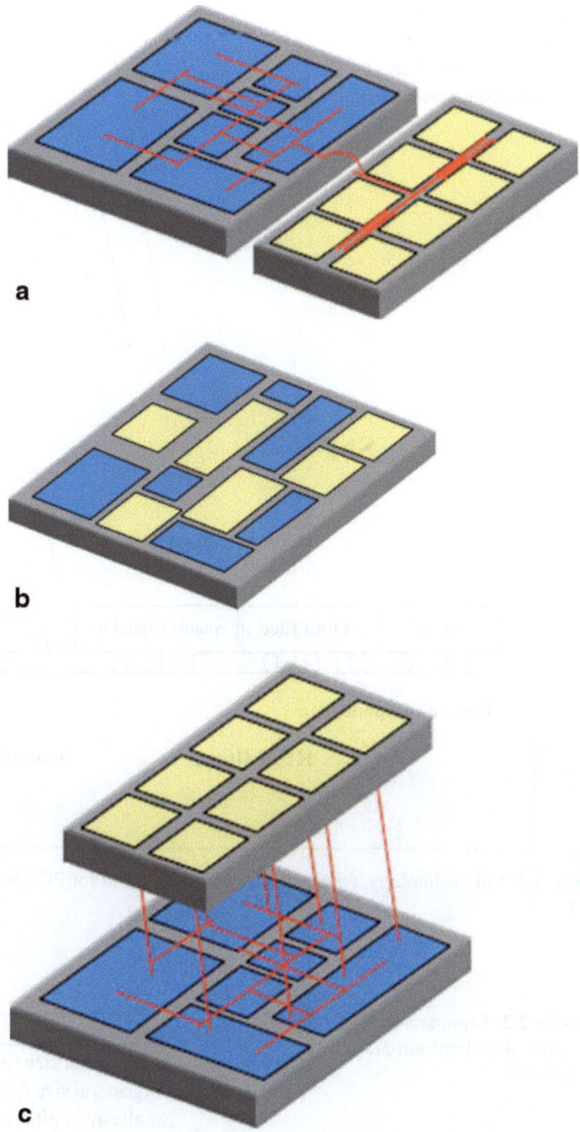

Fig. 2.1 A comparison between alternative design implementations: **a** single chips, **b** SoC, and **c** 3D [1]

lems by the use of short wires in 3D designs. The distance between layers is relatively small (~ 30–300 μm) in 3D-ICs compared to inter-router distance in 2D-ICs (~ 1000 μm or more) as is schematically shown in Figs. 2.4 and 2.5. These shorter wires will decrease the average load capacitance and resistance and decrease the number of repeaters which are needed to regenerate signal on long wires. More importantly, 3D technology enables the integration of heterogeneous technologies. In the 3D design, a 2D chip is divided into a number of different blocks, and

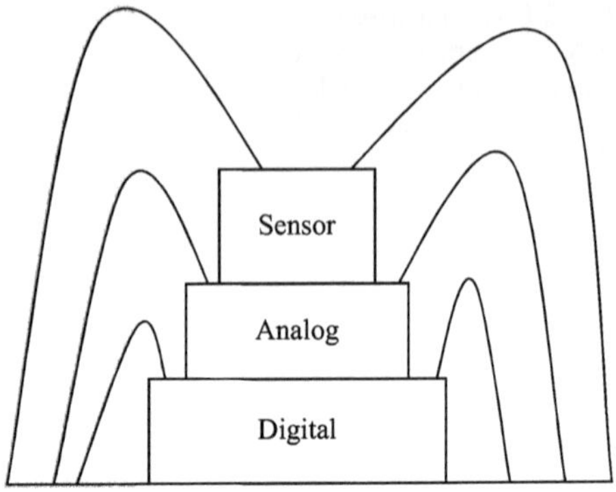

Fig. 2.2 SiP technology, where it can hire different chips/dies from different process in one package [1]

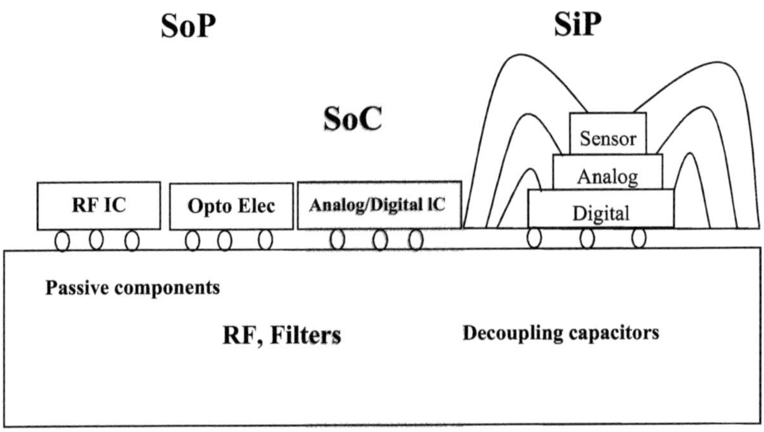

Fig. 2.3 SoP technology, where it is a "technical" term for PCB with bare dies in a single package [1]

Table 2.2 Comparison of feature sizes between 3D-IC and SoP

Feature	3D	SoP
Smallest via size (μm)	0.14	1
Largest via size (μm)	10	90
Smallest via pitch (μm)	0.4	10
Largest via pitch (μm)	200	200
Minimum silicon thickness (μm)	2	4
Maximum silicon thickness (μm)	70	300

each one is placed in a separate layer of silicon where each layer is stacked on top of each other, as schematically shown in Fig. 2.6. This may be exploited to build a SoC by placing different circuits with performance requirements in different lay-

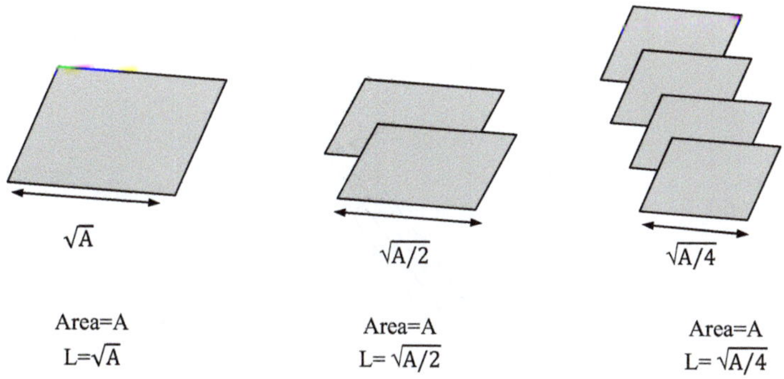

Fig. 2.4 Reduction in wire length where the original 2D circuit is implemented in two and four planes

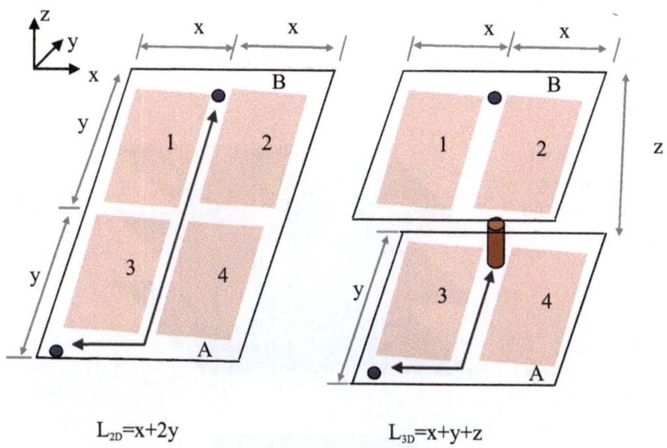

Fig. 2.5 Reduction in wire length where the original 2D circuit is implemented in two planes ($z<y$)

ers which enables novel "More than Moore" applications [1, 2]. Moreover, the 3D integration can be implemented on different levels: technology level, architecture level, or circuit level (Fig. 2.7). The advantages of 3D integration are summarized in Table 2.3.

In 3D integration, the communication among the stacked ICs and embedded components requires vertical interconnections. Since processing the vertical interconnections is challenging compared to the planar ones, the electrical and mechanical characteristics of the vertical interconnections can be a bottleneck for achieving the required system performance. Thus, various types of vertical interconnections are still being proposed. Several interconnection elements are already popular in industry. Figure 2.8 shows typical 3D structures with four types of interconnections,

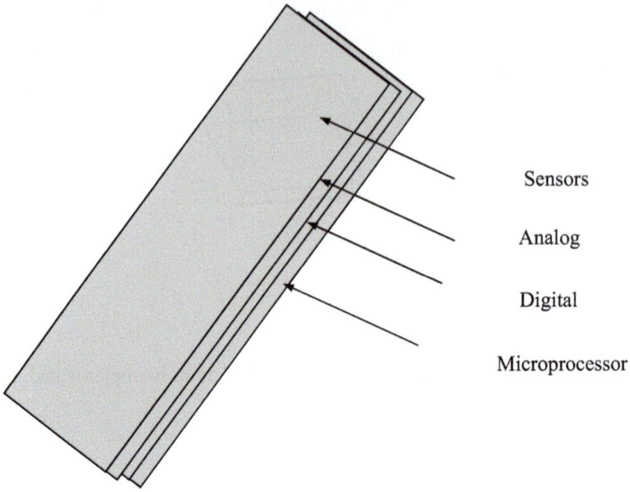

Fig. 2.6 A possible design for a 3D SoC (heterogeneous technologies)

Fig. 2.7 Different levels of 3D integration: technology level, architecture level, and circuit level [2]

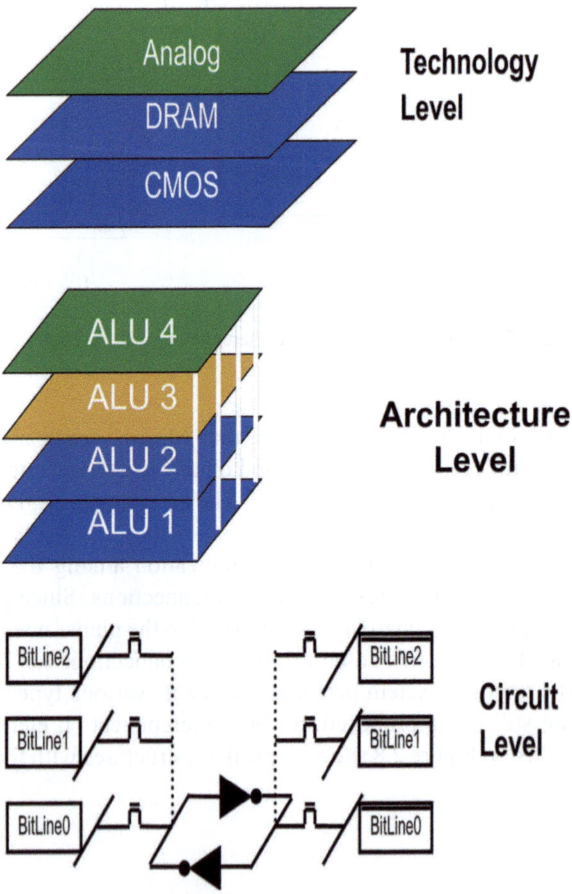

Table 2.3 Advantages of 3D integration

Advantages of 3D integration
1. Heterogeneous integration of devices
2. Reduced distance of global interconnects
3. Higher integration density
4. Higher form factor
5. Faster access between memory and logic device modules
6. Reduced overall resistance
7. Reduced power
8. Reduced packaging cost
9. Elimination of wire bonds
10. EMI and shielding
11. High bandwidth computer architectures
12. Use only one set of I/O pads to minimize parasitics due to I/O
13. Better reliability
14. Override the saturation of Moore's law

including bonding wires, TSV interconnections (which is a new silicon-based 3D packaging structure), metal bumps, and contactless communication (inductive and capacitive coupling). Table 2.4 summarizes the main difference between those four types of interconnects. It is clear that to improve electrical performance, TSV interconnections are becoming a major choice to replace the bonding wires and metal bumps [3]. TSV is not new, but now it is finally becoming a viable option [4].

2.2 TSV Structure

The general structure of a TSV (Fig. 2.9) is typically a cylinder with a uniform circular cross section of a conducting material (copper) surrounded by an insulator (silicon dioxide) which is intended to prevent leakage and resistive coupling through the substrate (silicon). 3D-ICs, which contain multiple layers of active devices, extensively utilize the vertical dimension to connect components.

Usually, the TSV interconnections are fabricated as an array in a silicon substrate. TSVs distribution is mainly determined during physical design such as partitioning, placement, and routing. TSVs can be laid out in a row, within a regular matrix, or within an irregular distribution depending on the applications and technology constraints [1–4].

The characteristics of a TSV are dependent on its geometrical parameters (such as the TSV diameter, height, pitch, and oxide layer thickness) and electrical parameters (such as the metal conductivity, the oxide permittivity, and the silicon substrate resistivity). The International Technology Roadmap for Semiconductors (ITRS) for the TSV is shown in Table 2.5.

The fundamental difference between TSVs and vias found in packages and PCBs is the semiconducting material surrounding them. Instead of looking at TSVs as

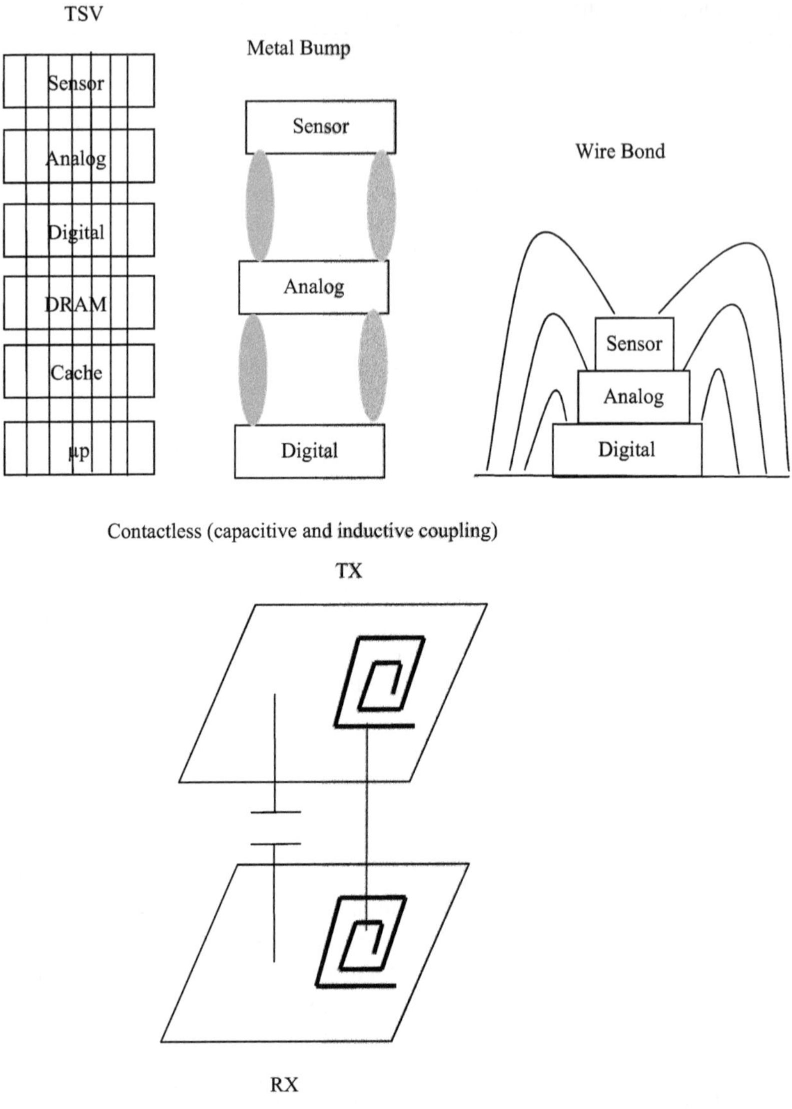

Fig. 2.8 Examples of 3D structures with interconnection elements, including bonding wires, TSV interconnections (which is a new silicon-based 3D packaging structure), metal bumps, and contactless communication (inductive and capacitive coupling)

wires, maybe other opportunities exist where the semiconducting properties of the substrate could be used to make the TSV function more like a device. With this thinking, the usefulness of the TSV can be enhanced and could potentially lead to enhanced performance and provide solutions for solving other problems being faced in 3D integration (the TSV can be considered to be a device, rather than using it only as a method of interconnect).

Table 2.4 Comparison among different 3D assembly technologies

Technologies	Advantages	Disadvantages
Wire bonding	Flexible connection High reliability Mature processing and cost-effective	Low density Long wiring Poor signal integrity Large pad area
Metal bumps	Short length	Large solder ball
TSVs	Short length High density Small footprint	Complex fabrication
Contactless (inductive and capacitive coupling)	Reducing the length of electrical paths Simple fabrication	Less reliability Cross talk and coupling issues Size of inductors (inductive coupling)

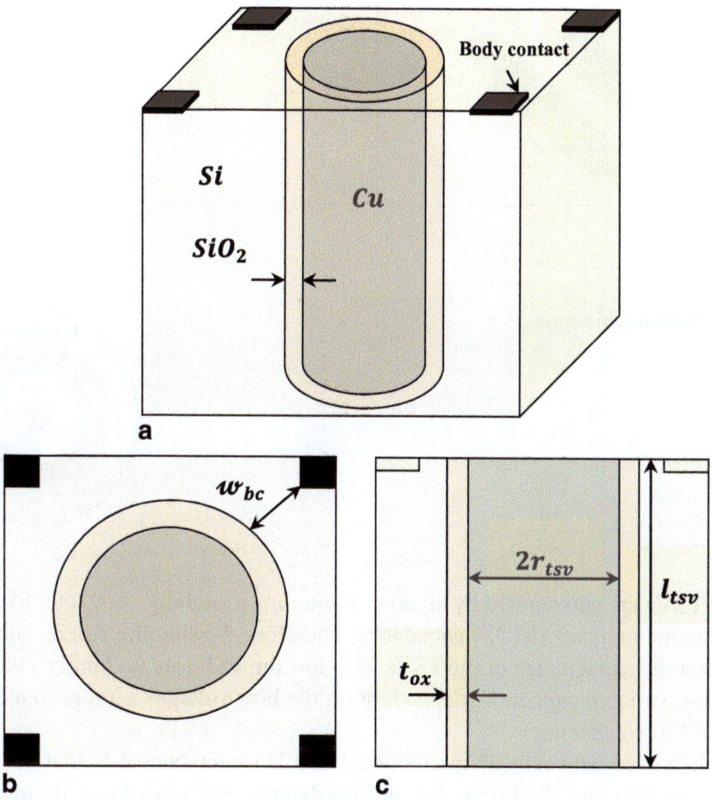

Fig. 2.9 The TSV structure assuming a uniform circular cross section of copper surrounded by an oxide inside a silicon substrate: **a** 3D view, **b** top view, and **c** side view

Table 2.5 ITRS for TSV

Parameters	2009–2012	2012–2015
Minimum TSV diameter (μm)	4–8	2–4
Minimum TSV pitch (μm)	18–16	4–8
Minimum TSV depth (μm)	20–50	20–50
Minimum TSV aspect ratio	1:10	1:20
Minimum contact pitch (μm)	20	10
Number of tiers	2–3	2–4

Table 2.6 3D and 2D wire model

3D wire model	2D wire model
Wires with finite width , thickness, and finite length 	Wires with finite width and thickness, but infinite length 

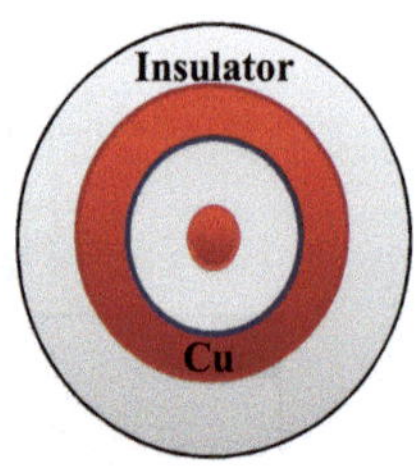

Fig. 2.10 Coaxial cable cross section

Since TSVs are surrounded by silicon, if you turn it laterally, they look like metal oxide semiconductor (MOS) capacitors. Therefore, biasing the silicon substrate should alter the capacitance of the TSVs. So, now the TSVs are no longer a wire but behave as a variable capacitor (depending on the bias voltage), similar to a device (such as a MOS capacitor).

The difference between 3D interconnect and 2D interconnect is that 2D wires have finite width and thickness, but infinite length. 3D wires have finite width, thickness, and finite length. These differences are depicted in Table 2.6. The shape can be rectangular or circular. But, from the simulation results, it can be found that the shape can affect the results very little. Moreover, a TSV is not like a coaxial cable which does not contain semiconductor as depicted in Fig. 2.10.

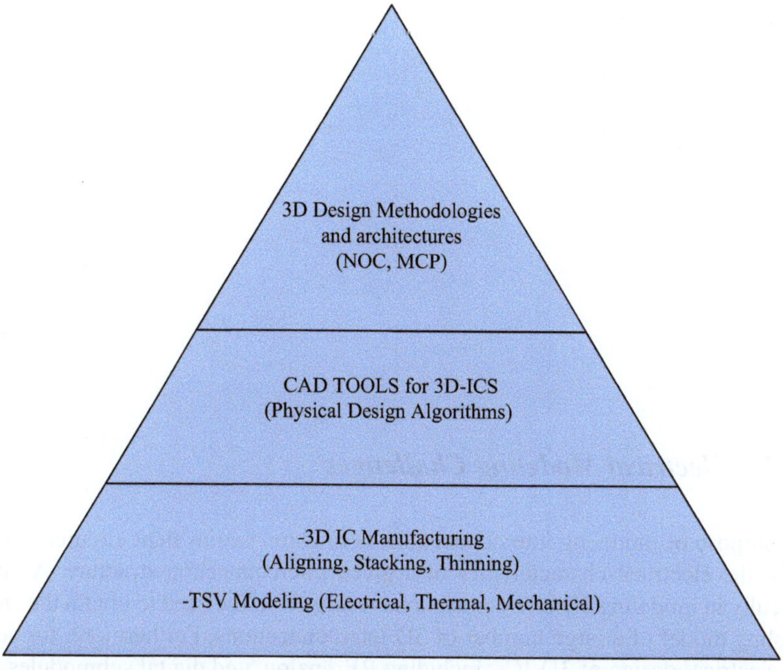

Fig. 2.11 3D-IC challenges: electrical and thermal modeling, 3D-CAD tools, yield and test, technological, and design methodology challenges

Table 2.7 3D challenges and their related work status

Challenge	Related work status
Electrical modeling	Research initiated; some models are available
Thermal modeling	Research initiated; some tools and techniques are available
3D-CAD tools	No complete solution is available; research needed
Yield and test	No solution is available; research is needed
Technological	Nearly mature
Design methodologies	No complete solution is available; research is needed
Circuit architecture	Research is needed

2.3 Modeling and Design Challenges for 3D-ICs

Although 3D-ICs offer many advantages over 2D-ICs, many challenges should be overcome before TSV-based 3D-ICs production becomes possible. These challenge levels are shown in Fig. 2.11 and discussed in the following subsection. Table 2.7 summarizes the 3D challenges and their related work status [5–9], where they include electrical and thermal modeling, 3D-CAD tools, yield and test, technological, and design methodology challenges.

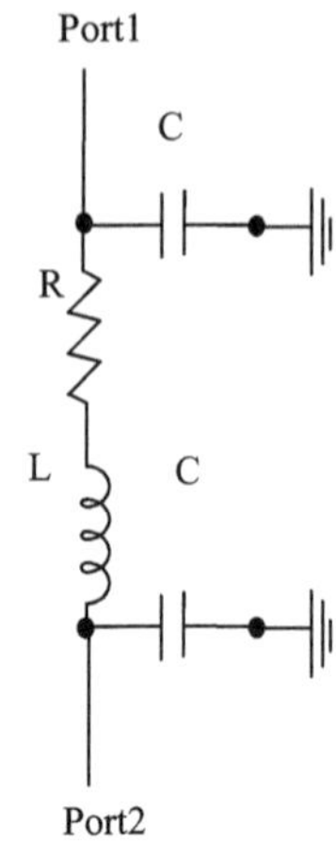

Fig. 2.12 A typical lumped model of a single interconnection (π model), which composed of a series inductor, a series resistor, and a parallel capacitor

2.3.1 Electrical Modeling Challenges

The purpose of modeling interconnections is to extract equivalent circuits that describe the electrical characteristics of a given interconnection structure. A major difficulty in modeling 3D interconnections results from the need to obtain the entire coupling model of a large number of 3D interconnections. Furthermore, for accurate electrical design of 3D-ICs, including RF, analog, and digital submodules, the 3D interconnection model should cover a sufficiently wide frequency range. The broadband model needs to capture frequency-dependent losses, coupling, and mismatch, which are contributed by the parasitic elements such as series inductances, resistances, shunt capacitances, and conductance. Unlike conventional interconnects, a TSV is surrounded by the silicon substrate that has a finite resistivity.

This finite resistivity results in a nonlinear capacitance and resistance parasitics that are not present in conventional interconnect, and a large capacitance that is highly dependent on TSV relative position to nearby body contact. Therefore, in case of modeling TSV interconnections, the silicon substrate should be considered as another source of loss. Clearly, TSV modeling is extremely challenging when a large number of TSV interconnections are involved. A typical lumped model of a single interconnection is composed of a series inductor, a series resistor, and a parallel capacitor (Fig. 2.12).

The values of circuit elements are determined by the shape of the interconnection itself as well as the coupling from nearby interconnections and other passive structures such as power/ground planes.

The TSV is a more general structure than transmission lines. Thus, applying transmission line theory for 3D interconnection modeling is not appropriate. Because of this fundamental difficulty with 3D interconnection modeling, most previous research relied on indirect ways for characterization based on measurements, analytical methods, and numerical methods.

The measurement-based equivalent model is difficult to extend to general multi-TSV structures due to the increased number of model parameters that need to be tuned to the multi-port measurement data. So the numerical methods are more commonly used [3].

One of the problems being researched in this domain is the modeling of multi-TSV arrangement in a 3D system and their time-domain co-simulation in a simulator like simulation program with integrated circuit emphasis (SPICE). Most of previous work characterizes specific TSV structures, but they do not address general multi-TSV problems [10–29]. Other related work proposes a model but no closed-form expressions are given for the TSV parasitic calculations [29]. In other words, there has been no related work in TSV macro-modeling. Also, few works address the multi-stacked TSV modeling problem.

2.3.2 Thermal Management Challenges

Thermal management is one of the important issues of 3D-IC integration. So, effective thermal management methodologies and solutions are needed for widespread use of 3D-IC integration. The thermal analysis is important to characterize the thermal stress distribution, and we can make layout modification and thus enhance the circuit reliability. The thermal issue of a 3D-IC is much severer than that of a 2D-IC. The cause is that the ambient environment of the die of a 2D-IC is the cooling material, but the ambient environment of a die within a 3D-IC may be another die that also generates heat. Therefore, thermal management in 3D-ICs is critical for maintaining the required reliability, performance, and power dissipation target. Therefore, new cooling techniques must be innovated to ensure that the chip can operate at appropriate temperatures and perform as required reliably [30, 31]. The recent techniques in thermal modeling and management are summarized in Figs. 2.13, 2.14, 2.15, 2.16, 2.17, 2.18.

2.3.3 CAD Tools Challenges

The addition of a third dimension would require support from more advanced CAD tools. 3D-IC physical design has attracted an increasing amount of attention. There is a significant amount of work on the floorplanning, placement, and routing for 3D-ICs. However, all these tools developed by different groups, using different formats to represent the design data, create barriers for researchers who need to make use of the existing design automation tools to conduct further studies on 3D-IC. Also, there is a lack of CAD algorithms, and there are only few commercially available electronic design automation (EDA) tools for 3D-ICs. Therefore, current 2D physical design tools (partitioning, placement, routing, timing, extraction, LVS, DRC) must enable 3D designs [32, 33]. One of the 3D-CAD design flows is shown in Fig. 2.19, where the main stages of 3D physical design cycle are [34–38]:

1. *Partitioning:* A chip may contain several million transistors. Due to the limitations of memory space and computation power available, it may not be possible

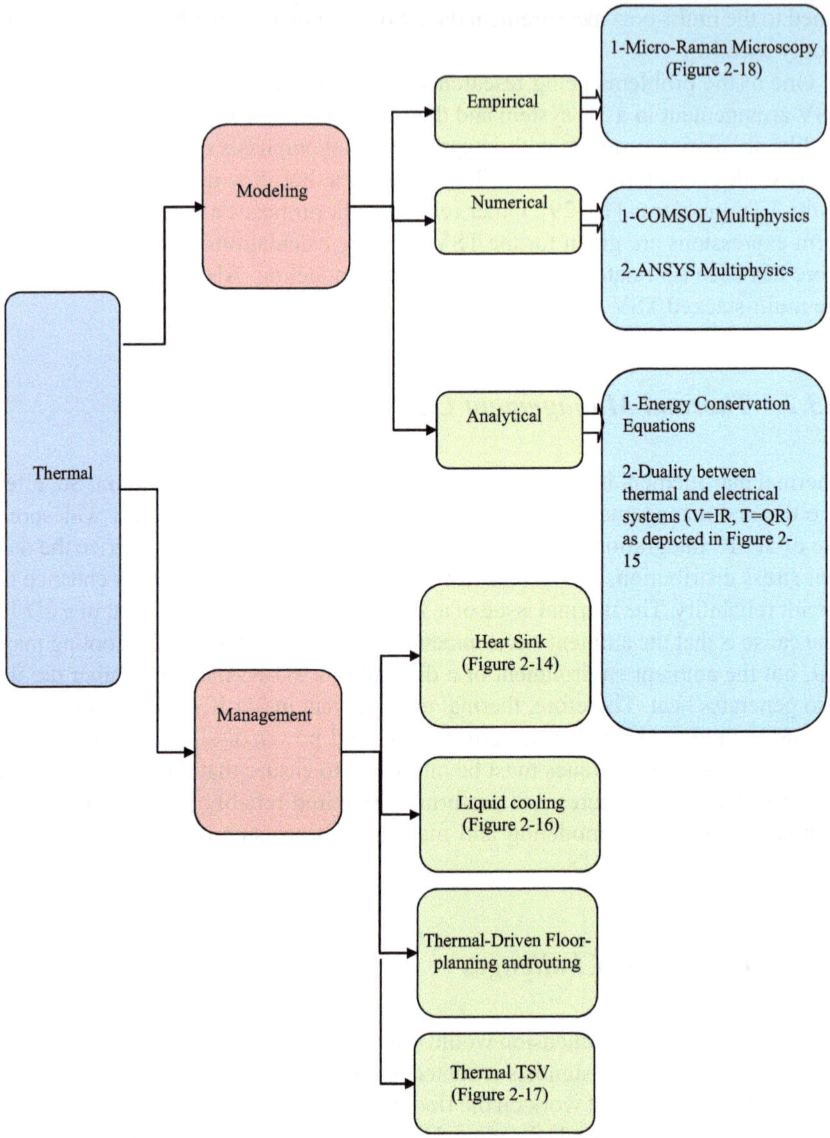

Fig. 2.13 Recent thermal modeling and management techniques

to lay out the entire chip (or generically speaking any large circuit) in the same step. Therefore, the chip (circuit) is normally partitioned into sub-chips (sub-circuits). These sub-partitions are called blocks. The actual partitioning process considers many factors such as the size of the blocks, number of blocks, and

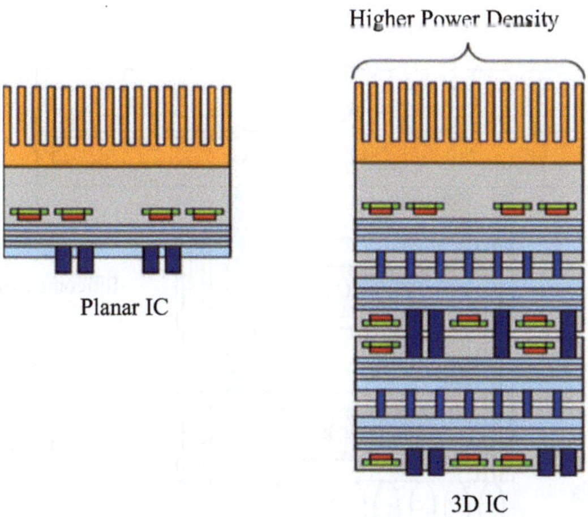

Fig. 2.14 Heat sink, where we have longer heat dissipation path in case of 4-die-stacked 3D-IC [30]

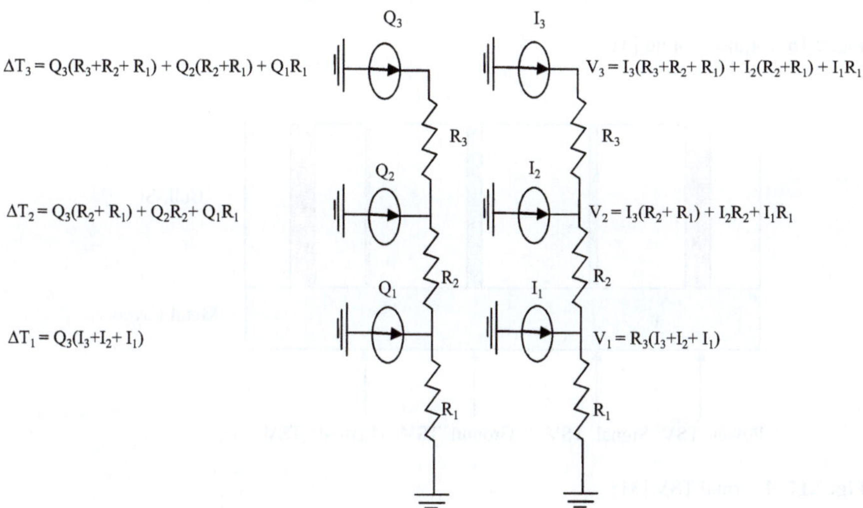

Fig. 2.15 Duality between thermal and electrical systems (*V=ualit=u*) [30]

number of interconnections between the blocks [39]. The output of partitioning is a set of blocks and the interconnections required between blocks. Figure 2.20 shows that the input circuit has been partitioned into three blocks. In large circuits, the partitioning process is hierarchical and at the topmost level a chip may have 5–25 blocks. Each block is then partitioned recursively into smaller blocks. There are two types of partitioning: timing-driven partitioning and thermal-driven partitioning.

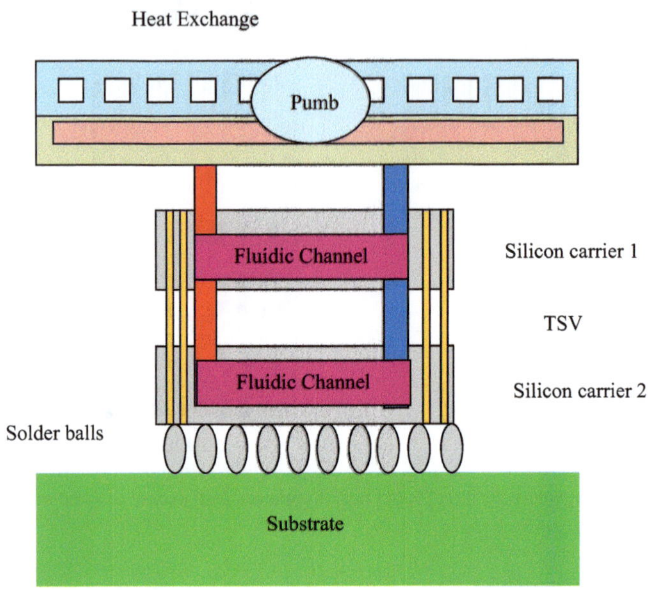

Fig. 2.16 Liquid cooling [31]

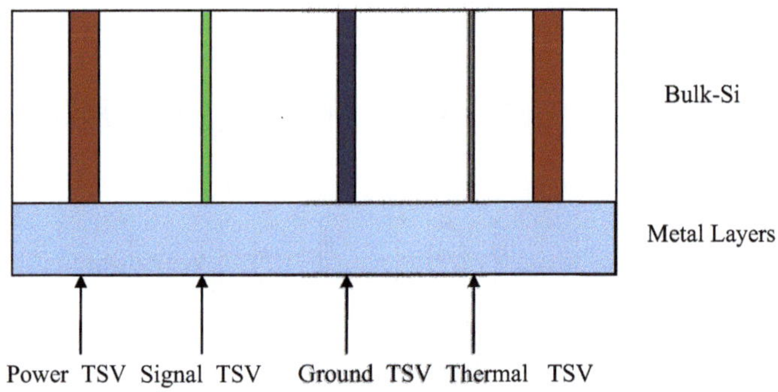

Fig. 2.17 Thermal TSV [31]

2. *TSV insertion:* Since TSVs occupy significant silicon area, a general tendency has been to use a minimum number of TSVs in 3D circuits. Moreover, determining the locations of TSVs is very important. The number of TSVs used is dependent on the number of 3D nets (nets that have pins in multiple dies), and the number of TSVs used for each layer [40]. Three kinds of TSVs exist in 3D designs: power and ground TSVs, signal TSVs, and clock TSVs. Signal TSVs and clock TSVs have relatively smaller diameter compared with power and ground TSVs. The inappropriate insertion of TSVs causes long routing path,

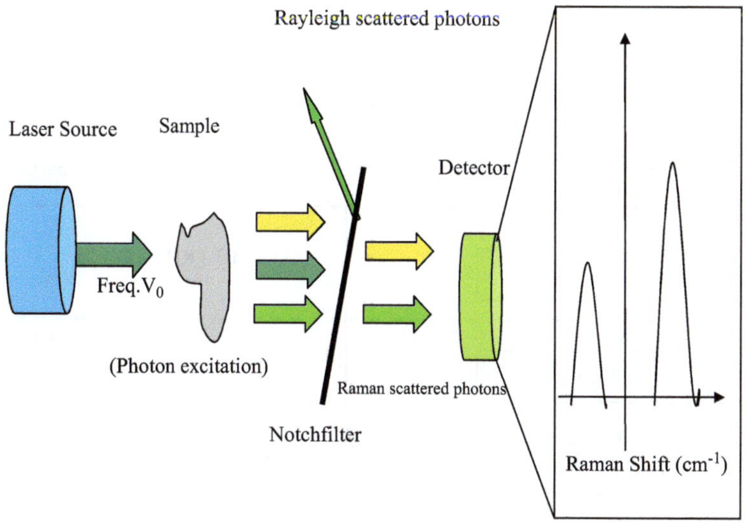

Fig. 2.18 Experiment setup for Raman measurements

increased chip area, or makes the performance of 3D-ICs worse than that of 2D-ICs.

3. *Floorplanning and placement:* This step is concerned with selecting good layout alternatives for each block, as well as the entire chip. The area of each block can be estimated after partitioning and is based approximately on the number and the type of components in that block. In addition, the interconnect area required within the block must be considered. The actual rectangular shape of the block, which is determined by the aspect ratio may, however, be varied within a prespecified range. Many blocks may have more general rectilinear shapes. Floorplanning is a critical step, as it sets up the groundwork for a good layout. However, it is computationally quite expensive. The goal of placement is to find a minimum area arrangement for the blocks that allows completion of interconnections between the blocks, while meeting the performance constraints. That is, we want to avoid a placement which is routable but does not allow certain nets to meet their timing goals. Placement is typically done in two phases. In the first phase, an initial placement is created. In the second phase, the initial placement is evaluated and iterative improvements are made until the layout has minimum area or best performance and conforms to design specifications. Figure 2.21 depicts that timing-driven floorplanning. In a nut shell, *Floorplanning:* details like shapes of blocks, I/O pin positions, etc., are not yet fixed (blocks with flexible shape are called soft blocks). *Placement:* Details like module shapes and I/O pin positions are fixed (blocks with no flexibility in shape are called hard blocks) [40].

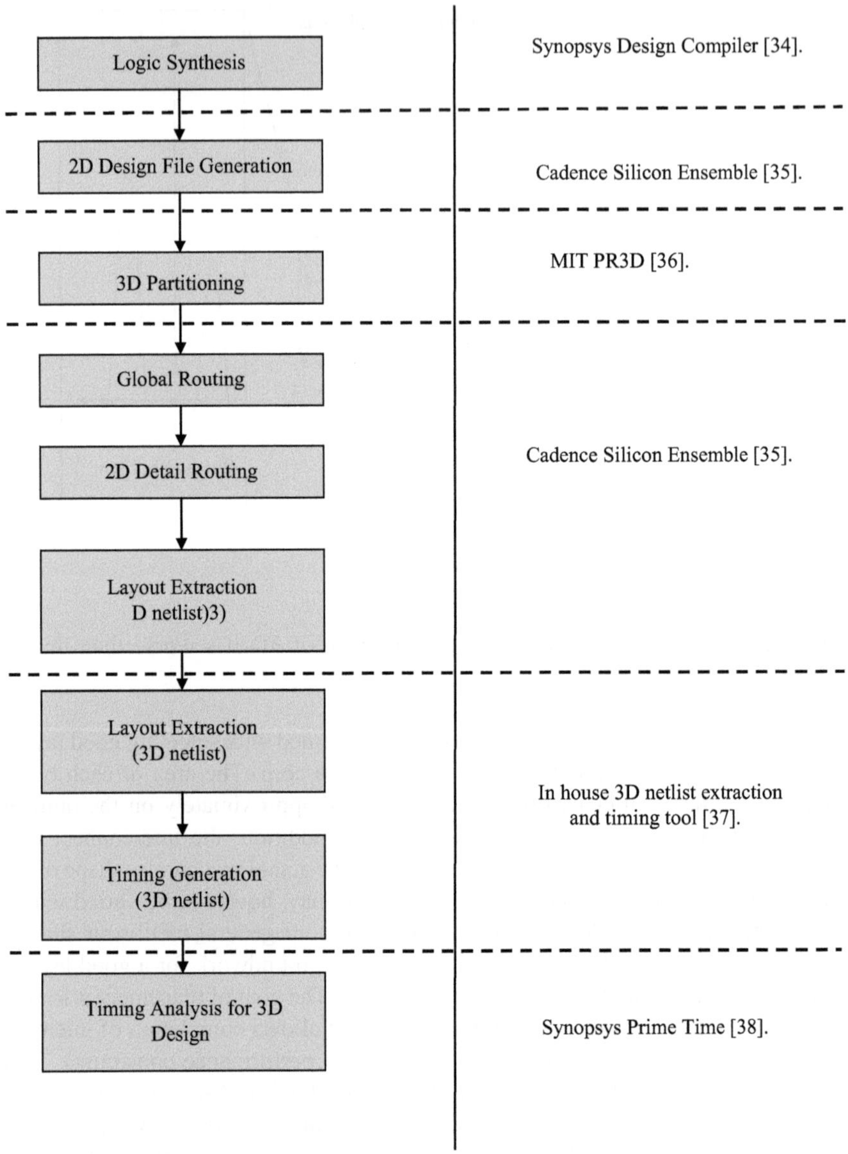

Fig. 2.19 Customized 3D-CAD design flow

4. *Routing:* The objective of the routing phase is to complete the interconnections between blocks according to the specified netlist (Fig. 2.22). The 3D routing process can be divided into global routing and detailed routing stages. Roughly speaking, the 3D global routing problems are very similar to the 2D routing problems, because the multiple metal layers already create some 3D structures. The additional considerations in 3D routing include (1) the solution space, which is larger than that of 2D routing because more layers are available; (2) the pins,

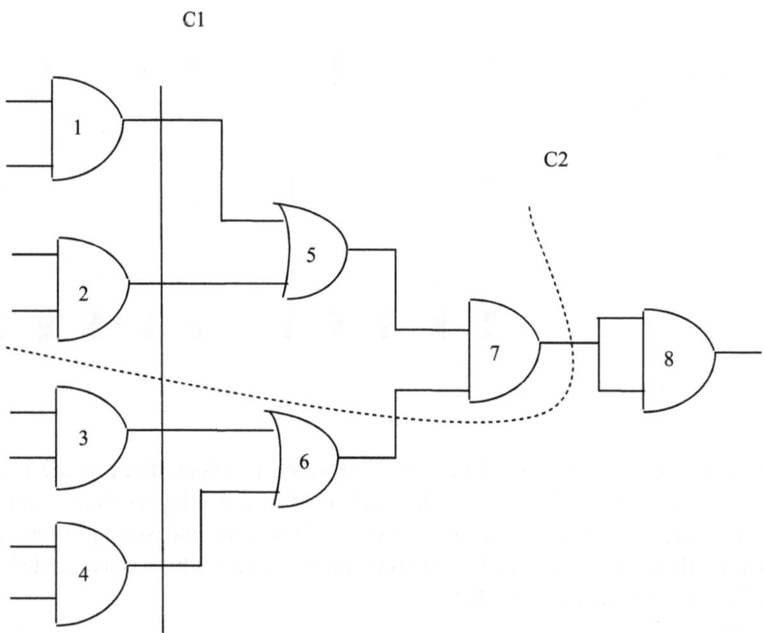

Fig. 2.20 A circuit to be partitioned; in partition using cutline C1, there are four off-chip wires, and in partition using cutline C2, there are two off-chip wires

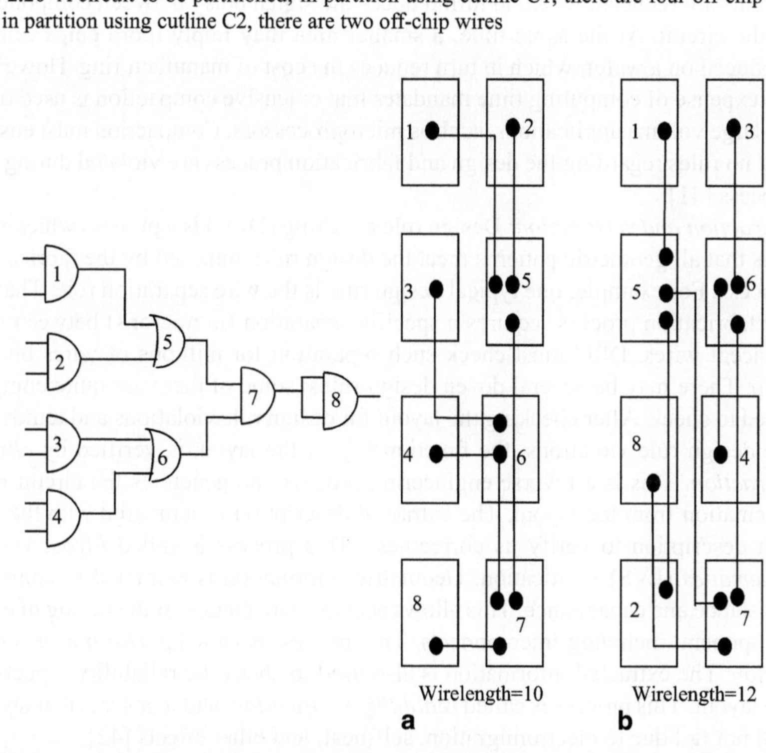

Fig. 2.21 Timing-driven floorplanning where in **a** floorplanning is routable and meets timing goals, **b** floorplanning is routable, but it does not meet timing goals

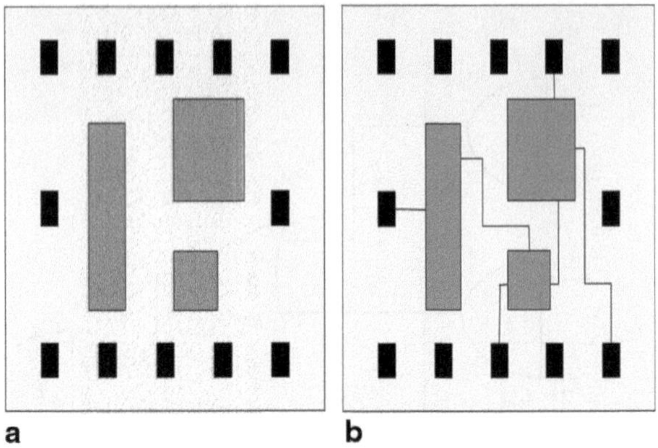

Fig. 2.22 Routing design cycle: **a** the blocks before routing and **b** the blocks after completion of the interconnections between blocks according to the specified netlist

which are located on more layers in 3D routing problems than in 2D routing problems, where the pins are only located on a few metal layers close to the silicon layer; (3) the TSV, which create blockages and consume routing resources to go through silicon layers; and (4) thermal optimization, which is available based on the idea of thermal TSV [41].

5. *Compaction:* Compaction is simply the task of compressing the layout in all directions such that the total area is reduced. By making the chip smaller, wire lengths are reduced, which in turn reduces the signal delay between components of the circuit. At the same time, a smaller area may imply more chips can be produced on a wafer, which in turn reduces the cost of manufacturing. However, the expense of computing time mandates that extensive compaction is used only for large volume applications, such as microprocessors. Compaction must ensure that no rules regarding the design and fabrication process are violated during the process [41].
6. *Extraction and verification:* Design rule checking (DRC) is a process which verifies that all geometric patterns meet the design rules imposed by the fabrication process. For example, one typical design rule is the wire separation rule. That is, the fabrication process requires a specific separation (in microns) between two adjacent wires. DRC must check such separation for millions of wires on the chip. There may be several dozen design rules; some of them are quite complicated to check. After checking the layout for design rule violations and removing the design rule violations, the functionality of the layout is verified by *circuit extraction*. This is a reverse engineering process and generates the circuit representation from the layout. The extracted description is compared with the circuit description to verify its correctness. This process is called *layout versus schematics* (LVS) verification. Geometric information is extracted to compute resistance and capacitance. This allows accurate calculation of the timing of each component, including interconnects. This process is called *performance verification*. The extracted information is also used to check the reliability aspects of the layout. This process is called *reliability verification,* and it ensures that layout will not fail due to electromigration, self-heat, and other effects [42].

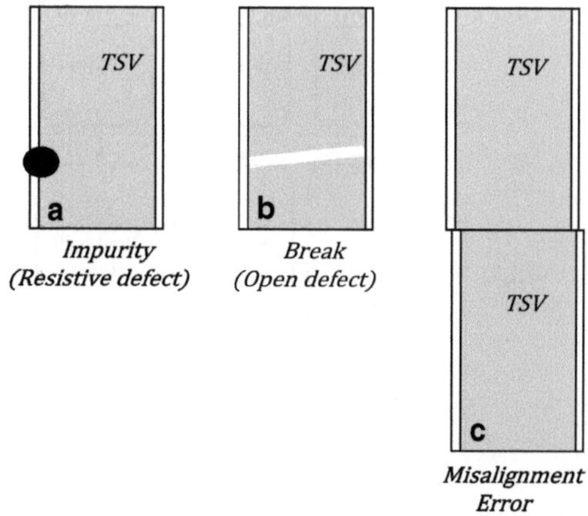

Fig. 2.23 **a** Schematic illustration of TSV shorts at the prebond stage (insulator short defect: leaking TSV), **b** conductor open defect (nonconducting TSV), equivalent open resistance, and **c** the misalignment defect of a TSV

Some of the available 3D CAD tools are:

1. *Layout editor tools*: 3D MAX, R3LOGIC.
2. *Physical design tools*: PSU/UCLA/NSU, MIT PR3D, MEVA-3D.
3. *Design analysis tools*: 3DCacti, HS3D to study the tradeoff between number of layers versus thermal and performance.
4. *Architecture-level analysis tools*: WormSim, 3D NoC Simulator. SunFloor 3D to explore design space.

2.3.4 Technological, Yield, and Test Challenges

There are new types of defects caused by TSV formation, and alignment (the alignment accuracy or bonding quality, including wafer-to-wafer alignment); a broad range of open issues are in a pressing need for novel ideas. The TSV-based 3D-IC fabrication process can result in several manufacturing defects, which causes 3D-IC failure. Filling high aspect ratio silicon vias with conductive metals is one of the most challenging problems in 3D integration. TSV-related defects might occur either in the fabrication of the TSVs themselves such as shorts between TSVs as shown in Fig. 2.23a, conductor open defect as shown in Fig. 2.23b, and in the bonding of the TSVs to the next tier as shown in Fig. 2.23c, or during the lifetime of the 3D such as mechanical load and thermal expansion. Solutions are needed to address these challenges before 3D integration technology can be widely used [43].

Also, it is well known that yield has a quadratic dependency on die size, and a linear dependence on chip count at a given die size. 3D designs may incur some yield loss due to vertical vias, and may gain some yield due to density [44]. There is still a lack of prebond test, postbond test, and DFT solutions [44, 45].

A conventional 2D test flow for planar wafers is depicted in Fig. 2.24a. Here, usually two test moments are applicable; i.e., a wafer test prior to packaging and a final test after packaging. During the manufacturing of 3D-ICs, additional test

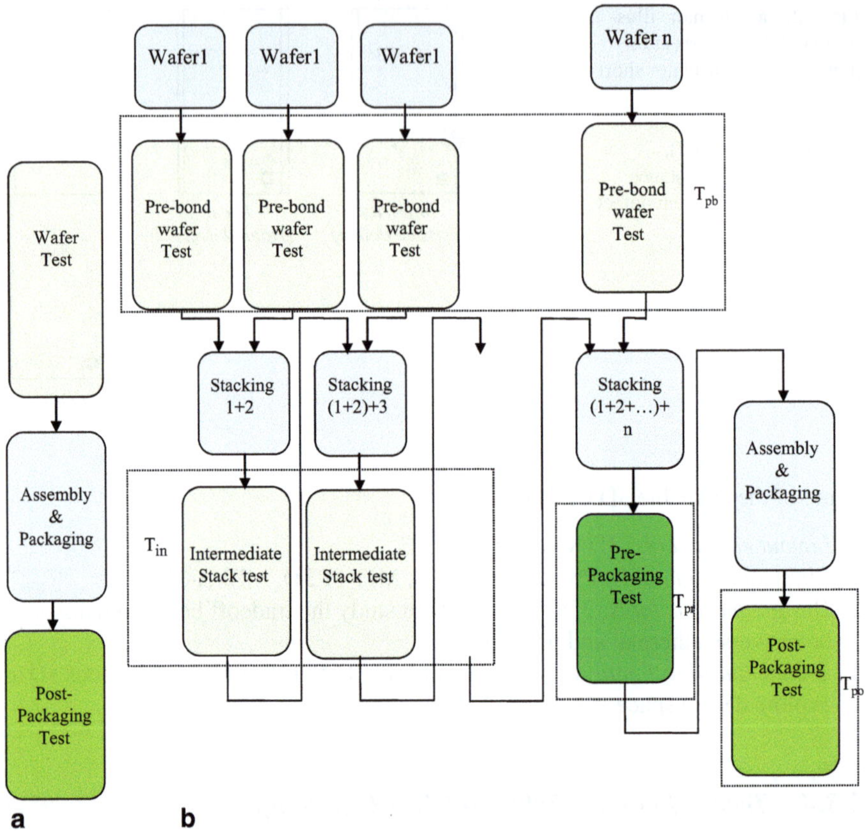

Fig. 2.24 **a** 2D versus **b** 3D D2W test flow framework. Note that in total there are $3n+1$ different test moments [43]

points can be defined for each partial created stack. At each test point, a distinction can be made between die tests and interconnect tests. Die tests ensure the functionality of individual dies, while interconnect tests ensure functional TSVs. For 3D-ICs, four test moments can be distinguished in time as depicted in Fig. 2.24b and explained next:

1. *Prebond (within-tier) wafer tests:* Since there are n layers to be stacked, we need n prebond wafer tests. It prevents faulty dies entering the stack. Besides die test, preliminary TSV interconnect tests can be applied (in case of via-first) as well. An example of a preliminary test that detects some faulty TSVs could be a capacitance test.
2. *Postbond (intermediate, tier-to-tier) stack tests:* The goal is to test portions of the system that could have been damaged during the bonding process; $n-1$ tests are applicable during the intermediate stacking and bonding. In this case, either the dies, the interconnects, their combination, or none of them can be tested for. Good tested dies in the prebond test phase could be corrupted during the stacking process as a consequence of, e.g., die thinning, or bonding. In the simulation

Table 2.8 Guidelines for design methodologies in TSV-based 3D-ICs

Problems	Solutions for signal integrity (SI), power integrity (PI), and thermal integrity (TI)
How to fill?	*HRS*: high-resistivity substrate [46]
How to size?	*Large SiO_2*: Increase the barrier oxide thickness during the process [18]
How to space?	*Increase pitch*: Based on the case studies and mathematical models, TSV pitch should be greater than 2.5 diameters [47]
How to drive?	*CM signaling*: Current mode signaling is the most effective one in alleviating coupled noise and increasing data rates [48, 49]
Others	*Shielding*: A solution consists in guarding the TSV using another TSV connected to the ground (both neighboring TSVs acts as shield, alleviate cross talk). This solution will be probably more efficient than a guard ring using body contact (less deep) [18]
	Body contacts: As TSVs introduce resistive and inductive voltage drop in 3D-ICs. This voltage drop at high frequency causes supply noise in 3D chips. The BCs make the inductive loop shorter [18]
	Thermal TSVs: Vertical vias are good thermal conductors They can be used as thermal vias to remove the heat from each die. As high temperature can affect the interconnect reliability [50]

model of our test flows, first the interconnects are tested and thereafter the dies in bottom-up order (in case both are tested for); if a fault is detected in the interconnects, then there is no need to test the dies as the 3D-IC will be faulty anyway. The reason for this particular order is that the test cost for interconnects is considered cheaper.

3. *Prepackaging test:* Only one test. This test can be applied after the complete stack is formed. Analogous to wafer testing in the 2D test flow, it can be applied to save unnecessary assembly and packaging costs. Both dies and interconnects can be tested for.
4. *Final postpackaging test:* Only one test. It can be applied after assembly and packaging to ensure the required quality of the complete 3D-IC. Other specific packaging-related tests could be applied here as well.

2.3.5 Design Methodology Challenges

3D design methodologies are still needed to reduce coupling between TSVs (body contacts reduces the current loop, which reduces the mutual inductances) and for clock and power distribution networks for 3D-ICs. Also, more investigations are required for analyzing the effect of TSVs on signal integrity, power integrity, and delay. Some guidelines for design methodologies in TSV-based 3D-ICs for signal integrity (cross talk noise, EMI, timing noise, jitter, and delay), power integrity (supply voltage noise, power delivery), and thermal integrity (reliability issue) are summarized in Table 2.8 and described below:

1. *Impact of LRS (low -resistivity substrate), MRS (medium), and HRS (high):* The insertion loss of the TSV in LRS is higher than in MRS and HRS (Fig. 2.25).

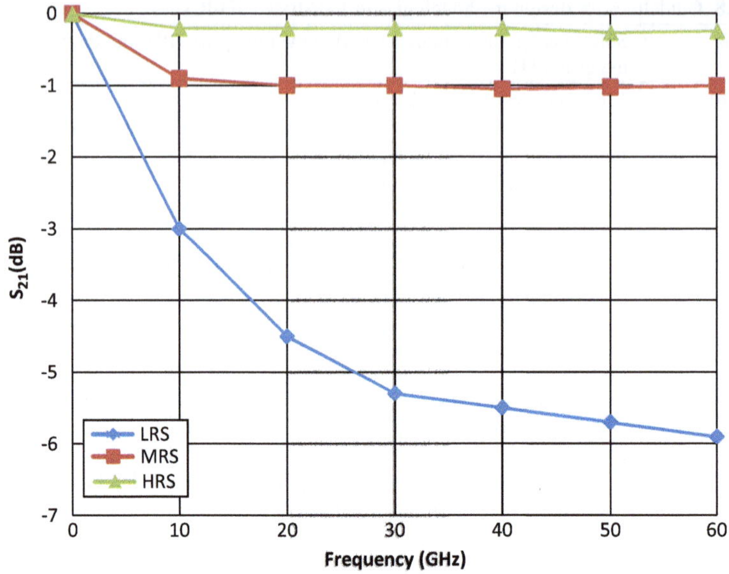

Fig. 2.25 Impact of low-resistivity substrate (*LRS*), high-resistivity substrate (*HRS*), and medium-resistivity substrate (*MRS*) on insertion loss

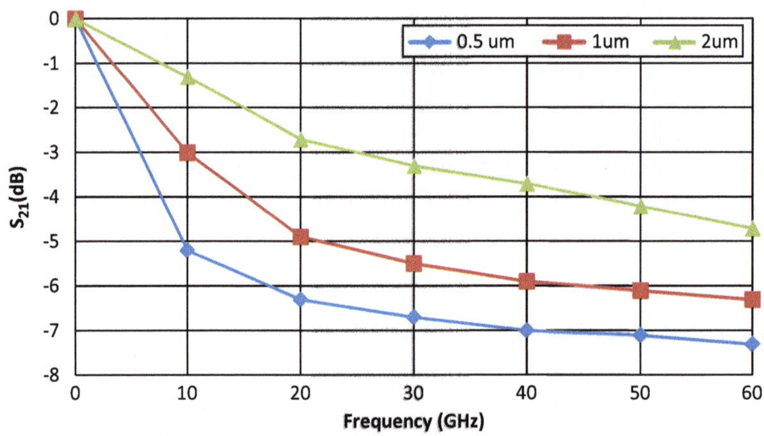

Fig. 2.26 Impact of thickness of SiO_2 on insertion loss

2. *Impact of thickness of* SiO_2*:* There is reduction in power loss when the thickness of SiO_2 is increased (Fig. 2.26).
3. *Impact of increasing pitch:* There is reduction in power loss when the pitch is increased (Fig. 2.27).
4. *Impact of signaling mode:* Current mode signaling is the most effective one in alleviating coupled noise and increasing data rates (Fig. 2.28).

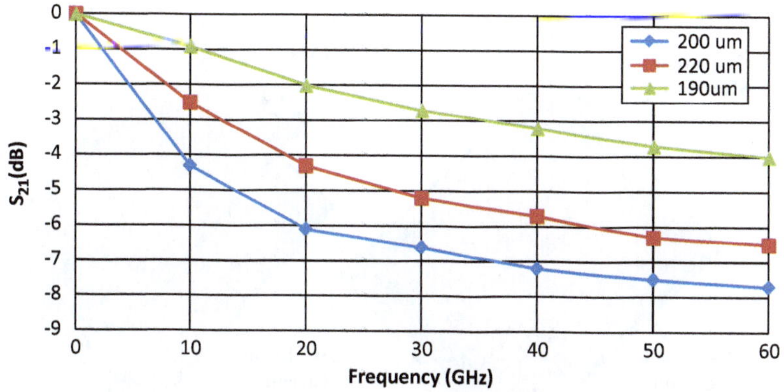

Fig. 2.27 Impact of increasing pitch between two TSVs on insertion loss

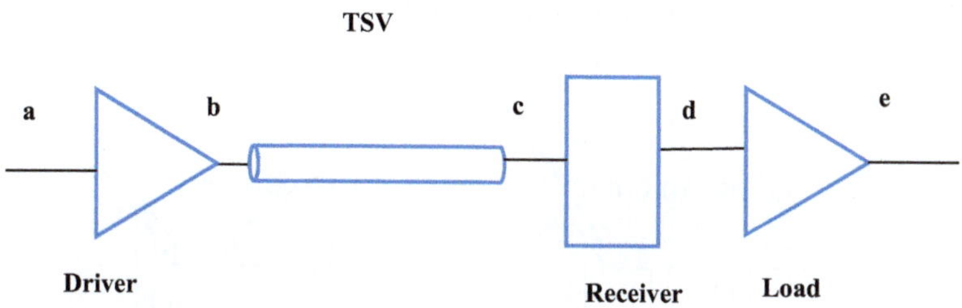

Fig. 2.28 Current mode signaling, where it is the most effective method in alleviating coupled noise and increasing data rates

5. *Impact of shielding:* Alleviates cross talk.
6. *Impact of body contact:* Alleviates cross talk.
7. *Impact of thermal TSVs:* Alleviating heat and improve thermal integrity.

2.3.6 Circuit Architecture Challenges

The 3D integration also provides opportunities for new circuit architectures for field-programmable gate arrays (FPGAs), memories, and others.

2.3.6.1 3D FPGA

A 2D FPGA requires 2D interconnections that grow faster than gate count. Thus, 2D FPGAs are limited in performance by the complexities of their internal interconnections. A 3D FPGA would overcome this interconnect limitations, resulting in faster data throughput, and faster switching of the gate-level configuration (Fig. 2.29) [1].

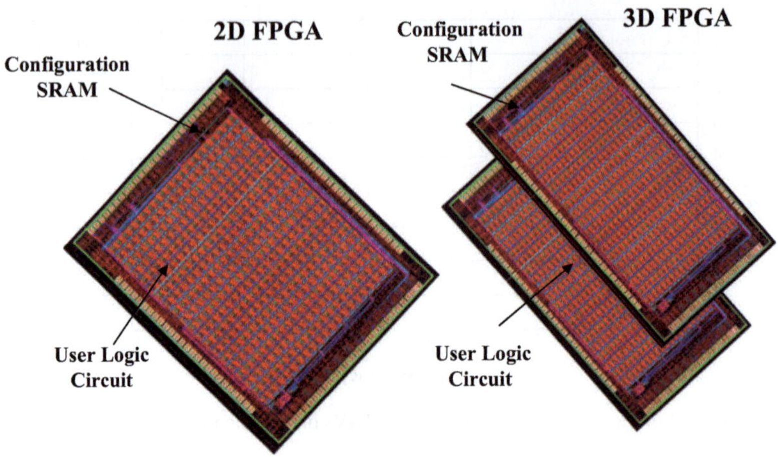

Fig. 2.29 3D FPGA structure as compared to 2D FPGA

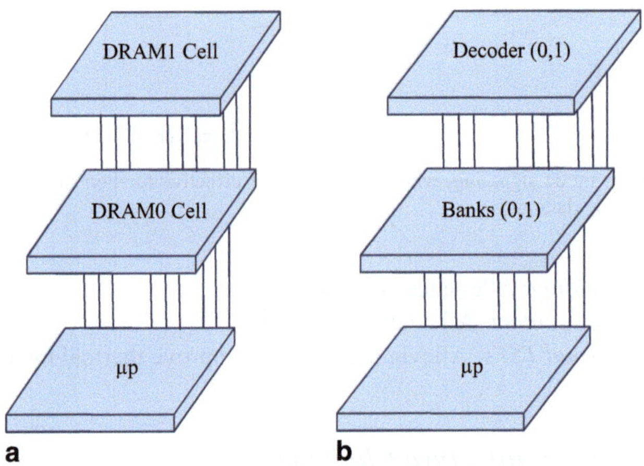

Fig. 2.30 **a** Complete memory in one layer and **b** split of memory elements in multiple layers

2.3.6.2 3D Memory

3D-stacked memories can be implemented in several architectures. One possible architecture is simply using TSVs to connect dynamic random-access memory (DRAM) layers to the processor layer as depicted in Fig. 2.30a. Such memory architecture reduces the long memory access latency, but does not provide much bandwidth benefits as the individual memory structures in each layer are still traditional 2D memory structures.

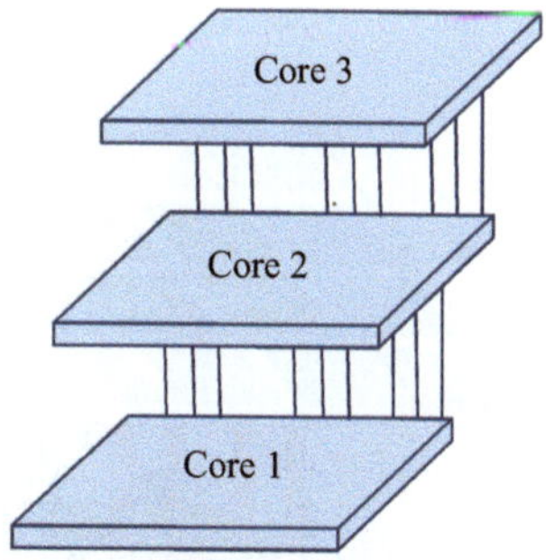

Fig. 2.31 3D microprocessor architecture, where each core is in a single layer [37]

DRAM architecture is introduced to fully exploit the benefits of 3D stacking, where individual storage-cell arrays are stacked in a 3D fashion. TSVs are used to link these memory arrays from different layers to peripheral logic (e.g., decoder and sense amplifier) in another layer as depicted in Fig. 2.30b. This is leading to significant increase of memory bandwidth.

2.3.6.3 3D Processor

3D can also be used to provide new capabilities to the microprocessor. 3D processors will be much more energy efficient than ordinary flat ones. Each core is in a single layer as depicted in Fig. 2.31 [37].

2.3.6.4 3D NoC

3D integration is mainly restricted by the communication infrastructure between different stacked dies of future multi-core SoC and NoC. NoC is a promising solution to simplify and optimize SoC design. NoC consists of switches and links and uses circuit or packet switching to transfer data through the system. NoC has been introduced as a shared communication medium that is highly scalable and can offer enough bandwidth to replace many traditional bus-based and point-to-point links (Fig. 2.32) [6].

2.3.7 Power Delivery and Clock Distribution Challenges

On-chip power supply noise has worsened in modern systems because scaling of the power supply network (PSN) impedance has not kept up with the increase in device density and operating current due to the limited wire resources and constant *RC* per wire length, and this situation is worsened in 3D-ICs.

In 3D-stacked ICs, about 30–40 % of the available die-to-die vias are used for power delivery and the rest are used for signals. Reducing the interconnect size with

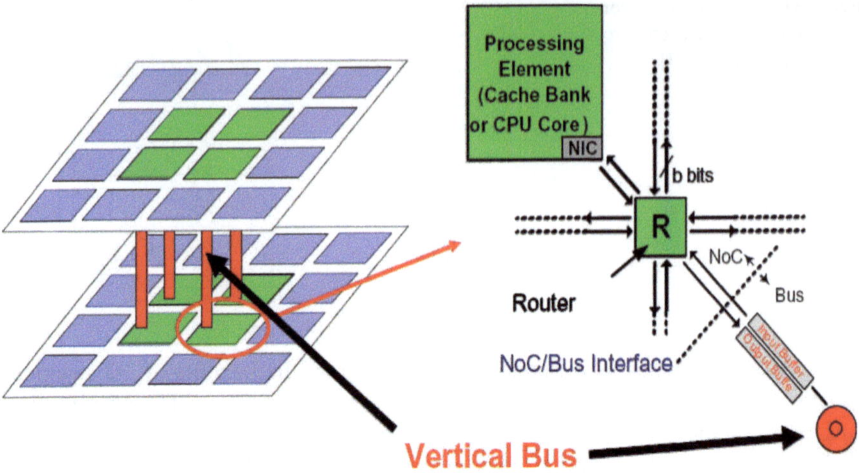

Fig. 2.32 3D NoC, where it consists of switches and links and uses circuit or packet switching to transfer data through the system [6]

each technology node and increasing speed with each generation increases *IR*-drop and *Ldi/dt* noise. The supply current flows through the inductive solder bumps and narrows TSVs with considerable parasitic inductances, resulting in a significant simultaneous switching noise (SSN). The 3D integration requires less than the original 2D design footprint and therefore the current density per pin is increased. As a result, the power distribution network in 3D systems needs to be accurately modeled and designed in order to satisfy the noise margins and power demand. Power is fed from the package through power I/O bumps distributed over the bottommost tier and travels to the upper tiers using TSVs.

The design of a clock distribution network in a digital IC is challenging in terms of obtaining low power consumption, low waveform degradation, low clock skew, and low SSN. In 3D-ICs, the effect of process variations on clock skew and jitter differs from 2D circuits. New schemes for a 3D clock distribution network with low skew, low jitter, low power consumption, and small area consumption are needed.

2.4 Summary

This chapter presented the detailed TSV structure as well as the challenges that should be overcome before the production of TSV-based 3D-ICs becomes possible, e.g., electrical modeling challenges, thermal and power challenges, technological challenges, design methodology challenges, and CAD tool development challenges.

References

1. K. Siozios, A. Bartzas, D. Soudris, Architecture-level exploration of alternative interconnection schemes targeting 3D FPGAs: a software-supported methodology. Int. J. Reconfig. Comput. **2008**, 8 (2008)
2. D.H. Kim, S. Mukhopadhyay, S.K. Lim, TSV aware interconnect length and power prediction for 3D stacked ICs, in *IEEE International Interconnect Technology Conference* (2009), pp. 26–28
3. K. Banerjee, S.J. Souri, P. Kapur, K.C. Saraswat, 3-d ics: a novel chip design for improving deep-submicrometer interconnect performance and systems-on-chip integration. Proc. IEEE **89**(5), 602–633 (2001)
4. K.P. Stuby, Hourglass-shaped conductive connection through semiconductor structures. U.S. Patent 3648131, Assignee: IBM, Filing Date: 7 Nov 1969, Issue Date: 7 March 1972
5. J.F. Li, C.W. Wu, Is 3D integration an opportunity or just a hype? in *ASPDAC*, 2010
6. L.P. Carloni, P. Pande, Y. Xie, Networks-on-chip in emerging interconnect paradigms: advantages and challenges, in *NOCs*, 2009
7. Y. Xie, J. Cong, S. Sapatnekar, *Three-Dimensional Integrated Circuit Design: EDA, Design and Microarchitectures* (Springer, Boston, 2010)
8. V.F. Pavlidis, E.G. Friedman, *Three-Dimensional Integrated Circuit Design* (Elsevier, 2011)
9. A. Papanikolaou, D. Soudris, R. Radojcic, *Three Dimensional System Integration* (Springer, New York, 2011)
10. C. Ryu, J. Kim, Electrical characterization of through silicon via (TSV) depending on structural and material parameters based on 3D full wave simulation, in *Electromagnetic Electronic Materials and Packaging*, 2007 Conference (May 2008), pp. 351–354
11. C. Ryu, J. Lee, H. Lee, K. Lee, T. Oh, J. Kim, High frequency electrical model of through wafer via for 3-D stacked chip packaging, in *Electronics System integration Technology Conf*, vol. 1 (Sept 2006), pp. 215–220
12. D.M. Jang, C. Ryu, K.Y. Lee, B.H. Cho, J. Kim, T.S. Oh, W.J. Lee, J. Yu, Development and evaluation of 3-D SiP with vertically interconnected through silicon vias (TSV), in *Proc. of ECTC* (2007), pp. 847–852
13. C. Ryu, J. Kim, J.S. Pak, Multi-stacked through-silicon-via effects on signal integrity and power integrity for application of 3-dimensional stacked-chip-package, in *IEEE*, 2009
14. S.W. Ho, S.W. Yoon, Q. Zhou, K. Pasad, V. Kripesh, J.H. Lau, High RF performance TSV silicon carrier for high frequency application, in *Electronic Components and Technology Conference*, 2008
15. A. Beece, K. Rose, T. Zhang, J. Qiang, Modeling and evaluation for electrical characteristics of through-strata-vias (TSVs) in three-dimensional integration, in *IEEE International Conference in 3D System Integration*, 2009, 3DIC 2009
16. J. Shim, E. Song, J. Pak, J. Lee, H. Lee, K. Park, J. Kim, Active circuit to through silicon via (TSV) noise coupling, in *Electrical Performance of Electronic Packaging and Systems, 2009. EPEPS '09*, IEEE 18th Conference
17. J. Kim, E. Song, J. Cho, J. Pak, J. Lee, H. Lee, K. Park, J. Kim, Through silicon via (TSV) equalizer, in *Electrical Performance of Electronic Packaging and Systems, 2009. EPEPS '09*, IEEE 18th Conference
18. M. Rousseau, O. Rozeau, G. Cibrario, G.L. Carval, M.A. Jaud, P. Leduc, A. Farcy, A. Marty, Through-silicon via based 3D IC technology: electrostatic simulations for design methodology, in *IMAPS Device Packaging Conference*, Phoenix, AZ, USA, 2008
19. I. Savidis, E.G. Friedman, Electrical modeling and characterization of 3-D vias, in *Proceedings of IEEE International Symposium on Circuits and Systems*, May 2008
20. I. Savidis, E.G. Friedman, Closed-form expressions of 3-D via resistance, inductance, and capacitance. IEEE Transac. Electr. Dev. **56**(9), 1873–1881 (2009)

21. R. Weerasekera, D. Pamunuwa, M. Grange, H. Tenhunen, L.-R. Zheng, Closed-form equations for through-silicon via (TSV) parasitics in 3-D integrated circuits (ICs), in *Proceedings of Workshop 3-D Integration, DATE Conference*, April 2009
22. E. Eid, T. Lacrevaz, S. de Rivaz, C. Bermond, B. Flchet, F. Calmon, C. Gontrand, A. Farcy, L. Cadix, P. Ancey, Predictive high frequency effects of substrate coupling in 3D integrated circuits stacking, in *3D System Integration, 2009*, 3DIC 2009
23. L. Cadix, A. Farcy, C. Bermond, C. Fuchs, P. Leduc, Modelling of through silicon via RF performance and impact on signal transmission in 3D integrated circuits, in *IEEE International Conference in 3D System Integration, 2009*, 3DIC 2009
24. T. Bandyopadhyay, R. Chatterjee, D. Chung, M. Swaminathan, R. Tummala, Electrical modeling of through silicon and package vias, in *IEEE International Conference in 3D System Integration, 2009*. 3DIC 2009
25. C. Xu, H. Li, R. Suaya, Compact AC modeling and analysis of Cu, W, and CNT based through-silicon vias (TSVs) in 3-D ICs, in *IEDM09-521*, 2009
26. M. Stucchi, K.D. Meyer, W. Dehaene, S.G. Katti, Electrical modeling and characterization of through silicon via for three-dimensional ICs. IEEE Transac. Electr. Dev. **57**(1) (2010)
27. D.E. Khali, Y.I. Ismail, M. Khellah, T. Karnik, V. De, Analytical model for the propagation delay of through silicon vias, in *Proceedings of IEEE International Symposium of Quality Electronic Design* (March 2008), pp. 553–556
28. I. Savidis, S. Alam, A. Jain, S. Pozder, R. Jones, R. Chatterjee, Electrical modeling and characterization of through-silicon vias (TSVs) for 3-D integrated circuits. Microelectr. J. **41**(1), 9–16 (2010)
29. R. Weerasekera, D. Pamunuwa, M. Grange, H. Tenhunen, L.-R. Zheng, Closed-form equations for through-silicon via (TSV) parasitics in 3-D integrated circuits, in *Proceedings of Workshop 3-D Integration, DATE Conference*, April 2009
30. P. Leduca, F.d. Crécya, M. Fayollea, B. Charleta, T. Enota, M. Zussya, B. Jonesb, J.C. Barbea, N. Kerneveza, S. Maitrejeana, D. Louisa, G. Passemardc, Challenges for 3D IC integration: bonding quality and thermal managementc, *IEEE*, 2007
31. T.Y. Chiang, S.J. Souri, C.O. Choi, K.C. Saraswat, Thermal analysis of heterogeneous 3-D ICs with various integration schemes, in *Proceedings of IEEE Electron Devices Meeting* (2001), pp. 681–684
32. C. Chiang, S. Sinha, The road to 3D EDA tool readiness, in *Proceedings of 2009 Conference on Asia and South Pacific Design Automation* (2009), pp. 429–436
33. A. Rahman, R. Reif, System-level performance evaluation of three-dimensional integrated circuits. IEEE Trans. VLSI Syst. (Special Issue on System-Level Interconnect Prediction) **8**(6), 671–678 (2000)
34. Synopsys Design Compiler, http://www.synopsis.com/
35. Cadence Silicon Ensemble, http://www.cadence.com/
36. MIT PR3D, http://mtlweb.mit.edu/researchgroups/icsystems/3dcsg/downloads.html
37. B. Vaidyanathan, W.-L. Hung, F. Wang, Y. Xie, V. Narayanan, M.J. Irwin, Architecting microprocessor components in 3D design space, www.cse.psu.edu/~yuanxie/Papers/VLSID07-3D.pdf
38. Synopsis PrimeTime, http://www.synopsis.com/
39. T. Yan, Q. Dong, Y. Takashima, Y. Kajitani, How does partitioning matter for 3D floorplanning? in *Proceedings of GLSVLSI* (April 2006), pp. 73–78
40. M. Pathak, Y. Joon Lee, T. Moon, S.K. Lim, Through silicon via management during 3D physical design: when to add and how many? in *IEEE International Conference on Computer-Aided Design (ICCAD)*, 2010
41. W. Hung, G. Link, Y. Xie, N. Vijaykrishnan, M.J. Irwin, Interconnect and thermal-aware floorplanning for 3D microprocessors, in *Proceedings of ISQED* (March 2006), pp. 98–104
42. Y.J. Lee, S.K. Lim, Routing optimization of multi-modal interconnects in 3D ICs, in *Proceedings of the 2009 Electronic Components and Technology Conference* (May 2009), pp. 32–39
43. K. Bernstein, P. Andry, J. Cann, P. Emma, D. Greenberg, W. Haensch, M. Ignatowski, S. Koester, J. Magerlein, R. Puri, A. Young, Interconnects in the third dimension: design chal-

lenges for 3D ICs, in *Proceedings of 44th Annual Conference on Design Automation* (2007), pp. 562–567

44. C. Ferri, S. Reda, I. Bahar, Parametric yield management for 3D ICs: models and strategies for improvement. ACM J. Emerg. Technolog. Comp. Syst. **4**, 22 (2008)
45. M. Hogan, D. Petranovic, Robust verification of 3D-ICs: pros, cons and recommendations, in *3D System Integration, 2009*, 3DIC 2009 (2009)
46. T. Bandyopadhyay, K. Han, D. Chung, R. Chatterjee, M. Swaminathan, R. Tummala, Rigorous electrical modeling of through silicon vias (TSVs) with MOS capacitance effects. IEEE Trans. Compon. Packag. Manuf. Technol. **1**(6), 893–903 (2011)
47. ALLVIA-The First Through-Silicon Via (TSV) Foundry, http://www.allvia.com/
48. H. Yu, J. Ho, L. He, Simultaneous power and thermal integrity driven via stapling in 3D ICs, Electrical Engineering Department, UCLA, 2009
49. K.C. Chillara, J. Jang, W.P. Burleson, Robust signaling techniques for through silicon via bundles, in *GLSVLSI'11*, May 2011
50. P.E. Garrou, E.J. Vardaman, P.D. Franzon, Through silicon via technology: the ultimate market for 3D interconnect, *Tech Search International*, Jan 2008

Chapter 3
TSV Modeling and Analysis

A robust wideband lumped element model for an arbitrary arrangement of multi through silicon vias (TSVs) is introduced. The introduced model takes into account the skin effects, the nonlinear TSV capacitance, coupling parasitics, adjacent body contact effects, as well as the parasitic elements due to the finite substrate resistivity. The introduced model consists of linear circuit elements that are directly compatible with simulation program with integrated circuit emphasis (SPICE) simulators, where a single nonlinear or frequency-dependent element is approximated by a number of linear, frequency-independent elements. The dimensional analysis method along with three-dimensional (3D) quasi-static electromagnetic field simulation is utilized to develop closed-form expressions for the lumped elements of the TSV model. This study also shows that the TSV capacitance is highly dependent on the placement and count of adjacent ground body contacts and has a value of tens of femtofarads in a typical current technology. This value is much higher than a minimum device capacitance and requires special design methodologies such as cascaded buffers.

3.1 Introduction

According to International Technology Roadmap for Semiconductor (ITRS), performance improvement of integrated circuits (ICs), beyond 45-nm technology, tends to saturate due to the fundamental physical limitation of the scaling and interconnect congestion bottleneck, unless a new paradigm shift is presented [1].

Three-dimensional IC (3D-IC) is now considered a new paradigm in IC design with a promising capability in overcoming the two-dimensional ICs (2D-ICs) interconnect congestion limitations, improving performance and allowing heterogeneous stacking of different processes such as radio frequency (RF), memory, logic, and microelectromechanical systems (MEMS) into a single 3D-IC [2].

Today, many 3D stacking technologies, such as through silicon via (TSV), wire bonding, metal bumping, and inductive coupling, are available, but TSV, shown in Fig. 3.1, has the advantage of introducing less delay, providing higher density, and

K. Salah et al., *Arbitrary Modeling of TSVs for 3D Integrated Circuits*,
Analog Circuits and Signal Processing, DOI 10.1007/978-3-319-07611-9_3,

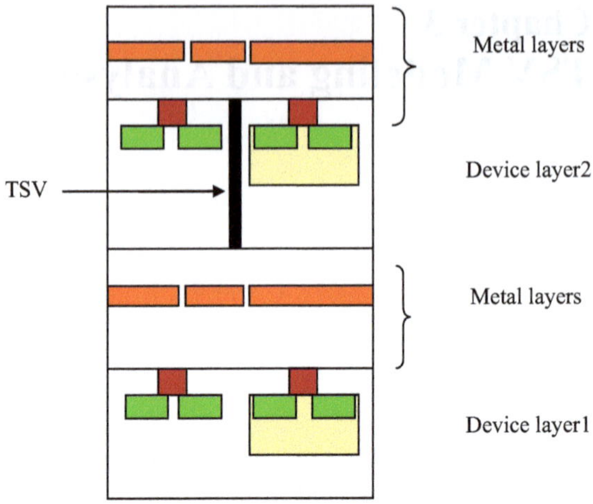

Fig. 3.1 Utilization of the TSV to connect components using the vertical dimension

having lower power feasibility than the other 3D stacking technologies as well as the on-chip wires in 45-nm technologies [3].

Therefore, developing a compact electrical model for the TSV and analyzing its impact on circuit performance is a topic that is receiving more attention. Developing such a model will find its use in system-conceptual-level explorations of 3D-ICs, and analyzing the TSV impact on circuit performance.

A typical structure of a TSV, as stated earlier, is a cylinder with a uniform circular cross section of a conducting material (copper) surrounded by an insulator (silicon dioxide) which is intended to prevent leakage and resistive coupling through the substrate (silicon).

Usually, TSVs are fabricated as an array in a silicon substrate as shown in Fig. 3.2 and can be laid out in a regular matrix or within an irregular distribution depending on the applications and technology constraints [4]. The characteristics of a TSV depend on its geometrical parameters (such as the TSV diameter, height, pitch, and oxide layer thickness) and electrical parameters (such as the metal conductivity, oxide permittivity, the silicon substrate resistivity, and the other surrounding structures around it). These characteristics are summarized in Table 3.1.

3.2 Modeling Methodology

The objective of this work, as stated earlier in the introduction, is to introduce a complete and robust model that accurately captures all the loss modes of a TSV, coupling parasitics between TSVs, and the TSV nonlinear capacitance and depletion region (MOS) effect. The methodology to achieve this objective can be summarized in the following steps (Fig. 3.3):

a. Derive a physics-based lumped element model for TSVs which captures the frequency dependency of the TSV using a set of lumped RLC elements.

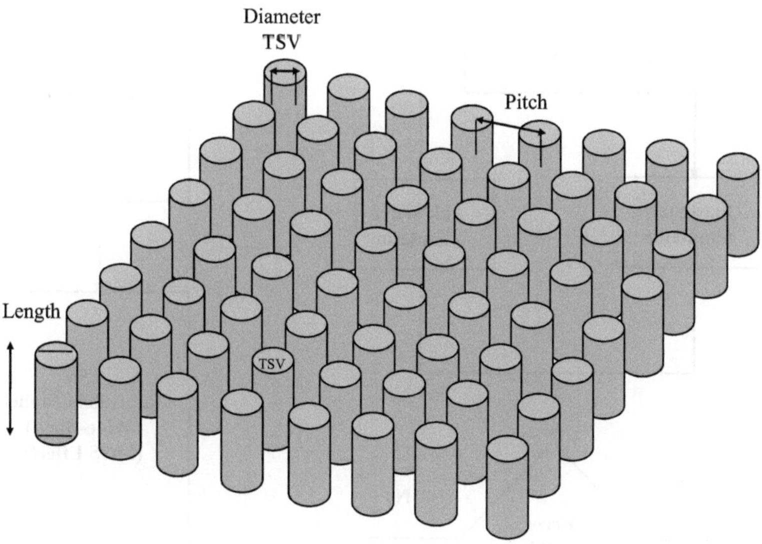

Fig. 3.2 TSV array configuration in a silicon substrate

Table 3.1 TSV physical parameters and their typical values

Geometrical parameters	Symbol	Typical current value (μm)
TSV radius	r_{tsv}	2.5
TSV height	l_{tsv}	Aspect ratio $* 2 * r_{tsv} = 50$
Dielectric thickness	t_{ox}	0.3
TSV-to-TSV separation	s_{tsv}	10
TSV pitch	p_{tsv}	$s_{tsv} + 2(t_{ox} + r_{tsv}) = 15.6$
Distance to body contact	w_{bc}	5
Number of body contacts	n_{bc}	4
Electrical parameters	Symbol	Typical current value
Conductor material	$f(\sigma_c, \mu_c, \varepsilon_c)$	$Cu = (58 \times 10^6, 1, 1)$
Dielectric material	$f(\sigma_{ox}, \mu_{ox}, \varepsilon_{ox})$	$SiO_2 = (0, 1, 4)$
Substrate	$f(\sigma_{si}, \mu_{si}, \varepsilon_{si})$	$Si = (10, 1, 12)$

b. Simulate the structure of the TSVs using electromagnetic (EM) simulation (either a quasi-static or a full-wave EM simulator).
c. Use the results of the EM simulator to optimize the physics-based lumped element values under a number of different setups (TSVs dimensions, TSV electrical and physical properties, and substrate parameters).
d. Use the dimensional analysis method [5] to develop closed-form expressions for the model lumped elements and use the results of step (d) to optimize the coefficients of the closed-form expressions.
e. Validate the model with its elements' value set using the closed-form expressions of step (d) against EM simulation.
f. Model linearization: represent the nonlinear elements with multiple linear elements.
g. Study the TSV impact on circuit performance.

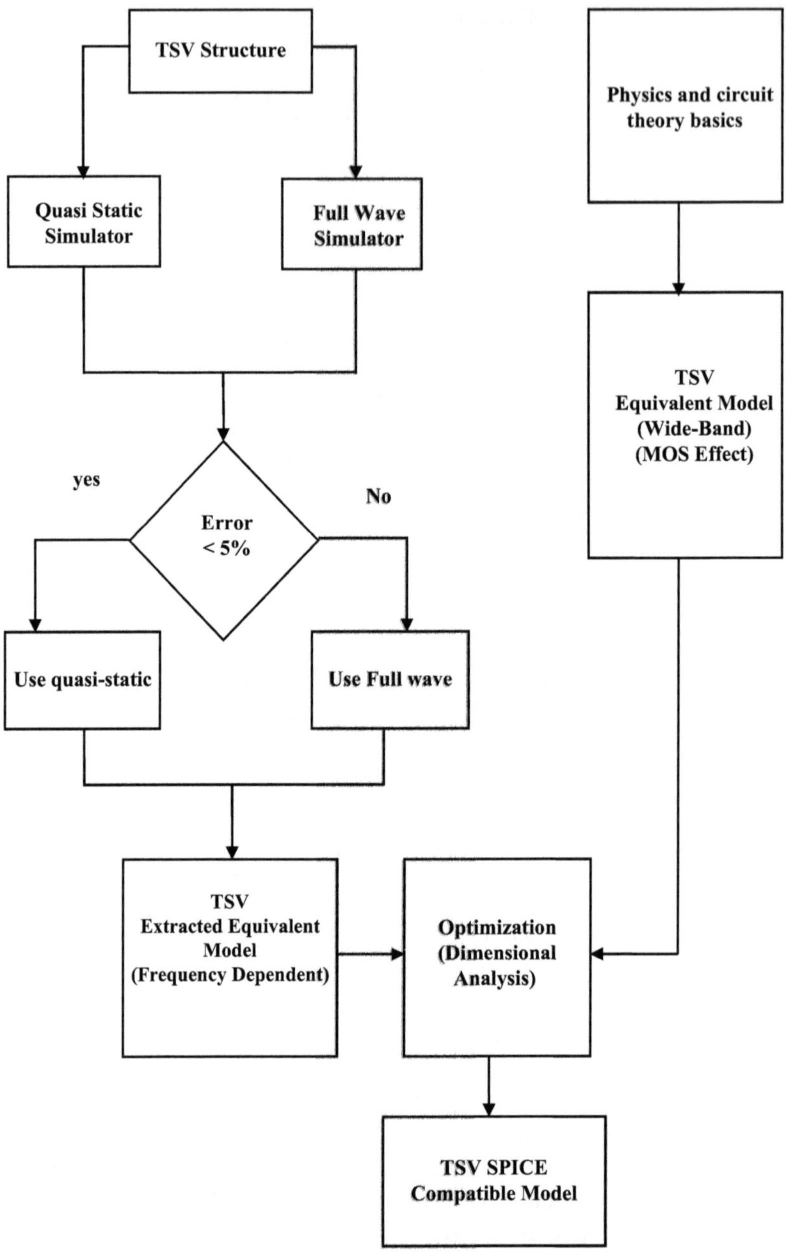

Fig. 3.3 The methodology for TSV modeling

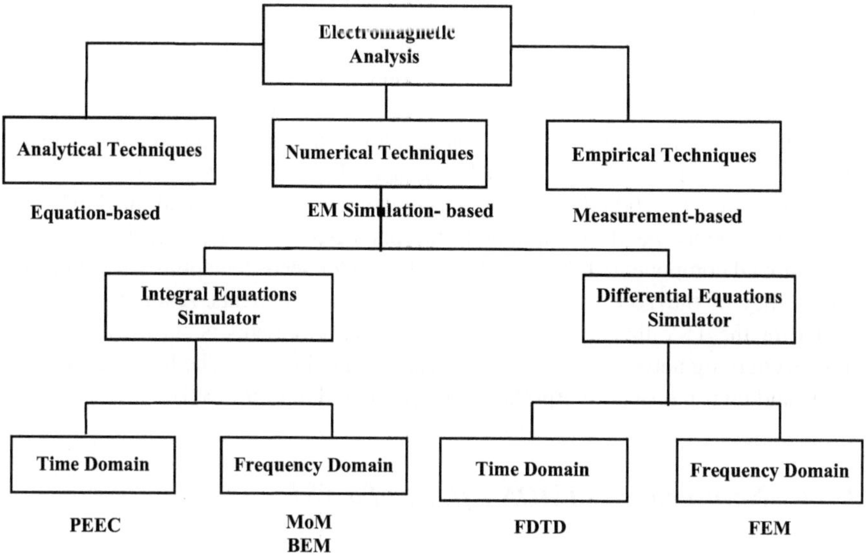

Fig. 3.4 Methods for solving Maxwell's equations

Table 3.2 Category of EM simulation-based techniques

	Time domain	Frequency domain
Integral equation-based	PEEC	MoM, BEM
Differential equation-based	FDTD	FEM

3.2.1 EM Modeling Methods

The modeling methods can be categorized as analytical, numerical, and empirical techniques (Fig. 3.4). The analytical techniques are equation-based, while the empirical ones are measurement-based. Regarding numerical techniques, they are EM simulation-based and classified into integral equation-based and differential equation-based [6]. Both of these EM simulation methods can be represented in time domain or in frequency domain (Table 3.2), and solved using partial element equivalent circuit (PEEC), method of moments (MoM), finite element method (FEM), boundary element method (BEM), or finite difference time domain (FDTD). In this work, the EM simulation-based technique will be used for TSV modeling.

3.2.2 Physics-Based Lumped Element Model for TSVs

An equivalent lumped RLC model of a TSV is required to efficiently represent its electrical performance in standard simulators. For the model to be accurate, it needs to capture the following TSV behaviors: conductor ohmic losses, skin effects, capacitance across the oxide, capacitance and resistance across the substrate, nonlinear capacitance due to metal oxide semiconductor (MOS) effect, and TSV–TSV interaction.

Stand-alone TSVs can be modeled using a lumped or a distributed circuit model. However, since the TSV size is much less than the wavelength of the considered frequency range (0–100 GHz), a single π lumped circuit model is adopted [7].

However, since nearly no simulator accepts a frequency-dependent description for a lumped component at the entry level, the model needs to capture the frequency dependency of the TSV using a set of lumped RLC elements. A lumped element model for the TSV that fulfills those previous requirements is shown in Fig. 3.5 and the coupling network between TSVs is shown in Fig. 3.6. The list of model elements and their physical interpretation are presented in (Table 3.3)..

3.2.3 Nonlinearities (MOS Effect) of a TSV

As stated earlier, the TSV model needs to capture the nonlinear capacitance due to the MOS effect. Hence, the physics of this phenomenon is studied. TSVs have a MOS-like capacitor structure, as illustrated in Fig. 3.7, where the TSV metal behaves similar to a gate and the silicon substrate behaves similar to the bulk of a MOS transistor [8–10]. This structure is described by the following set of equations:

$$r_T = r_{tsv} + t_{ox} + \omega_{dep} \tag{3.1}$$

$$r_1 = r_{tsv} + t_{ox} \tag{3.2}$$

$$r_0 = r_{tsv} \tag{3.3}$$

Where r_{tsv} is the TSV radius, t_{ox} is the oxide thickness, and w_{dep} is the substrate depletion width.

By solving Poisson's equation in a cylindrical coordinate system, the following equations result:

$$\frac{qN_a}{2\varepsilon_{si}}\left(r_T^2 ln\frac{r_T}{r_1} - \frac{r_T^2 - r_1^2}{2}\right) = 2V_t ln\frac{N_a}{n_i} \tag{3.4}$$

$$V_{th} = \phi_M - \chi - \frac{E_g}{2q} + V_t ln\frac{N_a}{n_i} + \frac{qN_a\left(r_T^2 - r_1^2\right)}{2\varepsilon_{ox}} ln\frac{r_1}{r_0}, \tag{3.5}$$

where

V_{th} is the threshold voltage,

ϕ_M is the metal work function,

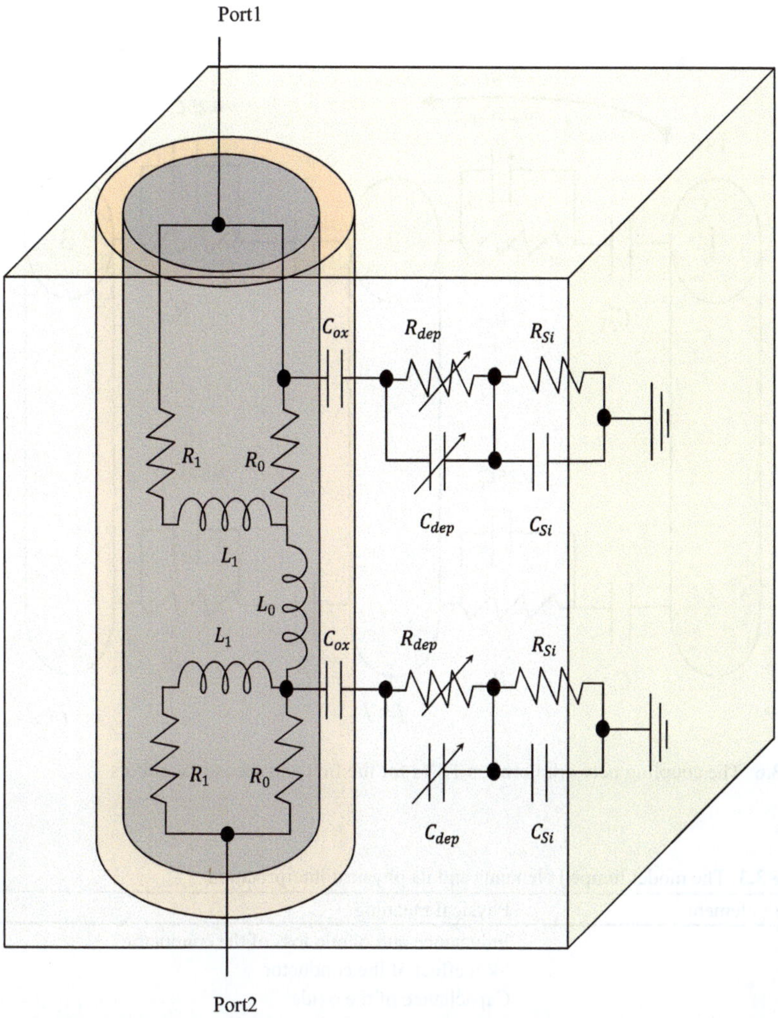

Fig. 3.5 The lumped model for a TSV based on a single π structure. The model is composed of R_0, L_0, R_1, L_1, C_{ox}, C_{si}, C_{dep}, R_{si}, and R_{dep}

χ is the electron affinity of Si (in V),
E_g is the band gap energy of Si (in eV),
q is the electronic charge $= 1.6 \times 10^{-19}$ coulombs,
$V_t = 0.026$ V (at 300 K),
N_a is the doping concentration of the acceptors ions, and
n_i is the intrinsic carrier concentration of Si (assuming p-type).

The full depletion approximation is used in deriving the above equations. This approximation assumes that there are no mobile charge carriers in the depletion region and the charge in this region is entirely due to the ionized atoms.

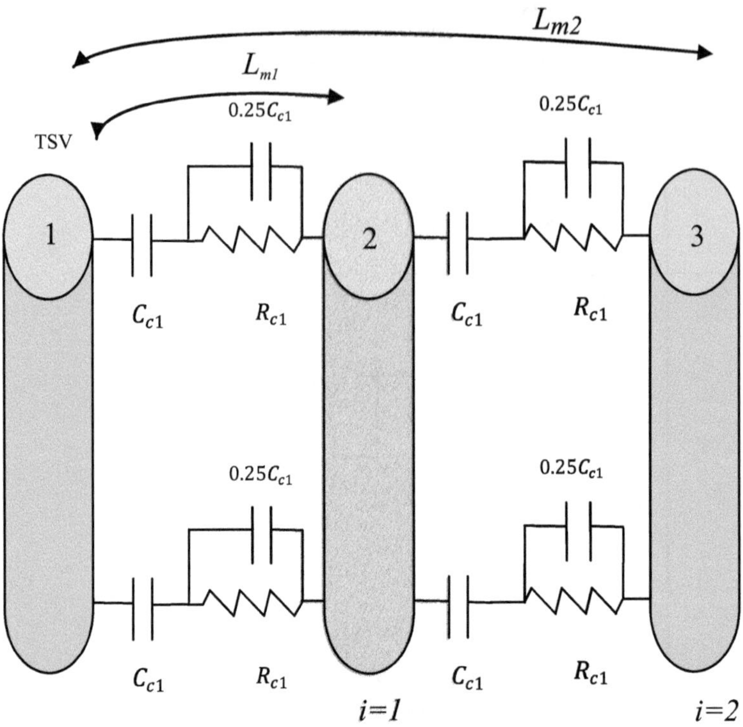

Fig. 3.6 The coupling network between TSVs for the first and second neighbors

Table 3.3 The model lumped elements and its physical interpretation

Circuit element	Physical meaning
R_0, L_0	Inductance and ohmic loss of the conductor
R_1, L_1	Skin effect of the conductor
C_{ox}	Capacitance of the oxide
C_{dep}, R_{dep}	Silicon substrate depletion region capacitance and resistance
C_{si}, R_{si}	Silicon substrate capacitance and resistance
C_c, R_c, L_m	Capacitive, resistive, and inductive coupling

By solving (3.4) and (3.5), a closed-form expression for the depletion width can be derived, which is given by

$$w_{dep} = -0.5r_1 + e^{-1.7r_1 W\left(\frac{-0.367qN_a r_1^2 - 0.8*10^7 * V_t \varepsilon_{si} ln\frac{N_a}{n_i}}{qN_a r_1^2}\right)}, \quad (3.6)$$

Fig. 3.7 Top view of a single TSV including the depletion region

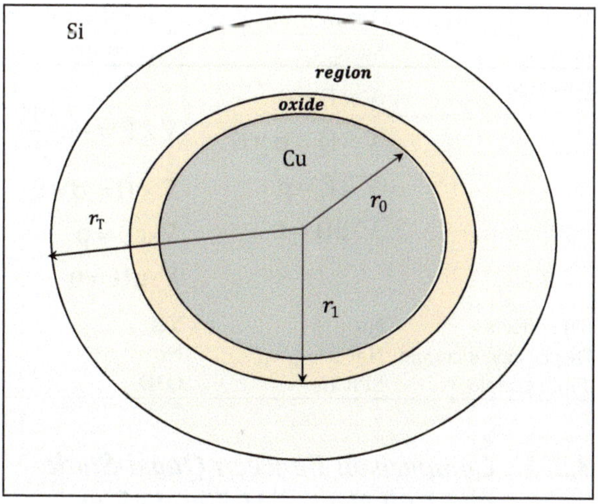

where W is the Lambert function and is calculated as follows:

$$W(z)=\sum_{n\geq 1}\frac{(-n)^{n-1}}{n!}z^{n} \tag{3.7}$$

By analogy to a MOS device [10], the TSV depletion capacitance dependency on V_{tsv} can be expressed as

$$C_{dep}=\frac{C_0}{\sqrt{1+\frac{V_{tsv}}{V_{th}}}} \tag{3.8}$$

where C_0 is the reference capacitance at $V_{tsv}=0$ V.

Along the same line, the duality concept between the resistance and the capacitance was introduced in the microwave area [11], and is used here to deduce the depletion resistance as

$$R_{dep}=\frac{\varepsilon\rho}{C_{dep}}=\frac{\varepsilon\rho\sqrt{1+\frac{V_{tsv}}{V_{th}}}}{C_0} \tag{3.9}$$

The closed-form expression for c_0 will be derived using the dimensional analysis method in Sect. 3.2.6.

Table 3.4 Comparison between different types of EM simulators

Item	Static	Quasi-static	Full-wave
Equations	$\nabla\times\vec{E}=0$ $\nabla\times\vec{H}=\sigma\times\vec{E}$ $\nabla\cdot\varepsilon\vec{E}=\rho$ $\nabla\cdot\mu\vec{H}=0$	$\nabla\times\vec{E}=-\mu\frac{\partial\vec{H}}{\partial t}$ $\nabla\times\vec{H}=\sigma\times\vec{E}$ $\nabla\cdot\varepsilon\vec{E}=\rho$ $\nabla\cdot\mu\vec{H}=0$	$\nabla\times\vec{E}=-\mu\frac{\partial\vec{H}}{\partial t}$ $\nabla\times\vec{H}=-\varepsilon\frac{\partial\vec{E}}{\partial t}+\sigma\times\vec{E}$ $\nabla\cdot\varepsilon\vec{E}=\rho$ $\nabla\cdot\mu\vec{H}=0$
Skin effects	No	Yes	Yes
Displacement current	No	No	Yes
Tool example	SPICE	Q3D	HFSS

3.2.4 Comparison Between Quasi-Static and Full-Wave EM Simulations

The methodology we introduced in this book, as detailed earlier, depends on simulating a structure of two adjacent TSVs and uses their EM simulation results to optimize the values of the lumped elements of the coupled TSV model. To conduct this step of the methodology, two different methods of EM simulation have been performed. The first method is based on a 3D full-wave field solver, high-frequency structural simulator (HFSS) [12], and the second one is based on a 3D quasi-static field solver, Q3D [13].

The EM simulators are classified into three types: static, quasi-static, and full-wave. The main difference in the static simulator is that both skin effect and displacement current are neglected, while in the quasi-static case, only the displacement current is neglected, and the full-wave simulator is the most accurate one. The EM simulator does not take into consideration the carrier transport equation, i.e., the nonlinearities of semiconductors, while the device simulator is semiconductors-physics-based, but it does not take into account magnetic field effects. Moreover, EM simulators neglect quantum effects.

Table 3.4 summarizes the main differences between these two types of simulators and a static simulator, for completeness. Figure 3.8 shows a comparison between the full-wave and the quasi-static results for the S_{21} parameter of a TSV in a frequency sweep from few Hertz to 100 GHz, where the depletion region is simulated as a silicon region with conductivity equal to zero and width equal to w_{dep}.

The comparison shows a 5 % difference between the two methods where the difference is calculated using log function. The error is deemed acceptable; hence, the quasi-static simulator is used in the rest of this book.

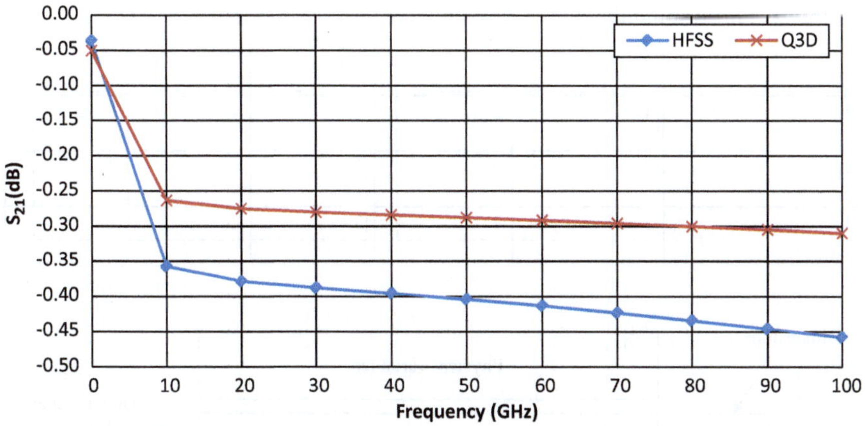

Fig. 3.8 Comparison between the S_{21} for a TSV produced by a full-wave simulator (HFSS) and a quasi-static simulator (Q3D; $(2r_{tsv} = 5\,\mu m,\ l_{tsv} = 50\,\mu m)$)

Table 3.5 A dimensional matrix example

	V_1	V_2	V_3	V_4	V_5
d_1	1	2	3	4	5
d_2	2	4	3	0	2
d_3	3	4	3	2	1

3.2.5 Dimensional Analysis Theory

Dimensional analysis is one of the most powerful modeling methods where it is used to reduce the number of physical variables through the identification of the variables' dimensions and combining the original variables to deduce a canonical number of dimensionless variables that makes curve fitting to obtain final equations simpler. In a physical relation, the dimensions d_1, d_2 ... of each variable V_1, V_2, ... are best given and displayed by the dimensional matrix in which the rows are the dimensions and the columns are the variables. A typical element of this matrix is the exponent to which the particular dimension (row) is raised in the particular variable (column). Table 3.5 shows a dimensional matrix in which the elements are of course fictitious, as they are chosen for illustration only. For example, the dimension of variable V_2 is $d_1^2 \ldots d_2^4 \ldots d_3^4$ and of V_4 is $d_1^4 \ldots d_2^0 \ldots d_3^2 = d_1^4 \ldots d_3^2$.

Partition the dimensional matrix into submatrices A and B, where A should be square and nonsingular whose order is the number of dimensions $N_d = 3$ as depicted in Table 3.6. Matrix A must not be singular; i.e., its determinant must not be zero. A is an $N_d \times N_d$ square matrix and B is an $N_d \times (N_v - N_d)$ matrix. Different products of variables in a particular relationship called π_1, π_2, ... π_n. Number of π is the order of matrix B. Referring to Fig. 3.9, π terms, the dimensionless parameters,

Table 3.6 Partitioning the dimensional matrix into submatrices A and B

	V_1	V_2	V_3	V_4	V_5
d_1	1	2 [B]	3	4	5 [A]
d_2	2	4	3	0	2
d_3	3	4	3	2	1

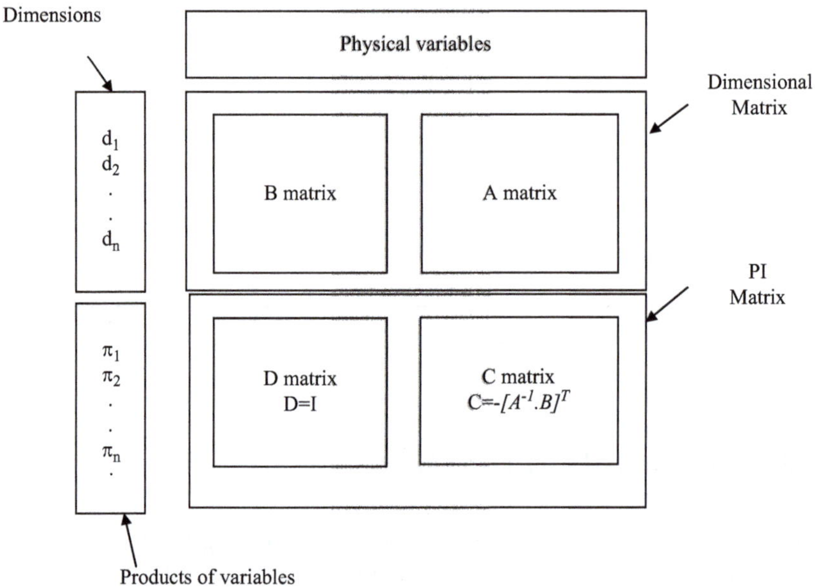

Fig. 3.9 TSV's structure of the dimensional set

are calculated from C and D matrices, where $C = -[A^{-1} \cdot B]^T$, $D = I$, and the relation between different π terms is given by $\pi_1 = f(\pi_2, \pi_3, \ldots \pi_n)$.

3.2.6 Closed-Form Expressions for TSV Model Elements Using Dimensional Analysis

Dimensional analysis is applied as a new design methodology for TSV modeling. By using the dimensional analysis and the curve fitting technique, simple formulas of TSV parameters, such as resistance, capacitance, and inductance, are obtained.

The closed-form equations for the TSV are extracted in two steps. The first step is fitting the field solver-based *S*-parameters up to 100 GHz to the equivalent lumped element model using dimensional analysis and curve fitting techniques. The

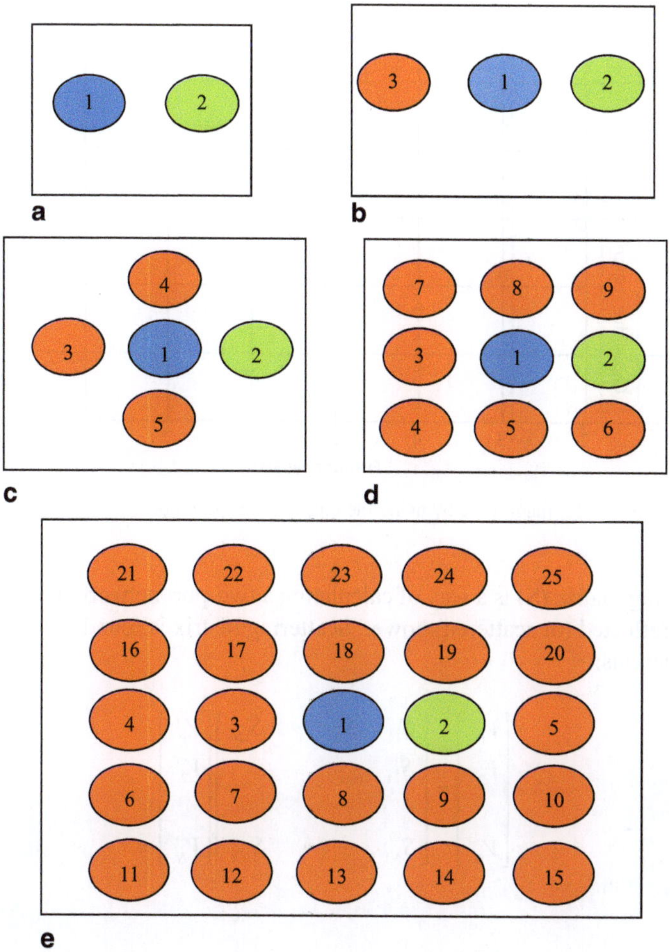

Fig. 3.10 TSVs structures under test: **a** two straightly arranged TSVs, **b** three straightly arranged TSVs, **c** five arranged TSVs, **d** 3 × 3 matrix TSVs, and **e** 5 × 5 matrix TSVs. All the TSVs are signal TSVs

second step is using forms suggested by the physical laws to guide the curve fitting process to produce the closed-form equations with the help of the dimensional analysis method.

There are two common methods of curve fitting: polynomial curve fitting and least squares curve fitting. In polynomial curve fitting, a polynomial is found for which each of the data points occurs on the polynomial curve. In least squares fitting, a polynomial is found for which the sum of the squares of the errors between the data points and the polynomial is minimal.

The data sets' input to the dimensional analysis method and curve fitting will be obtained from EM simulations for different TSV arrangements. Therefore, the TSV parasitics are investigated here for different structures (Fig. 3.10).

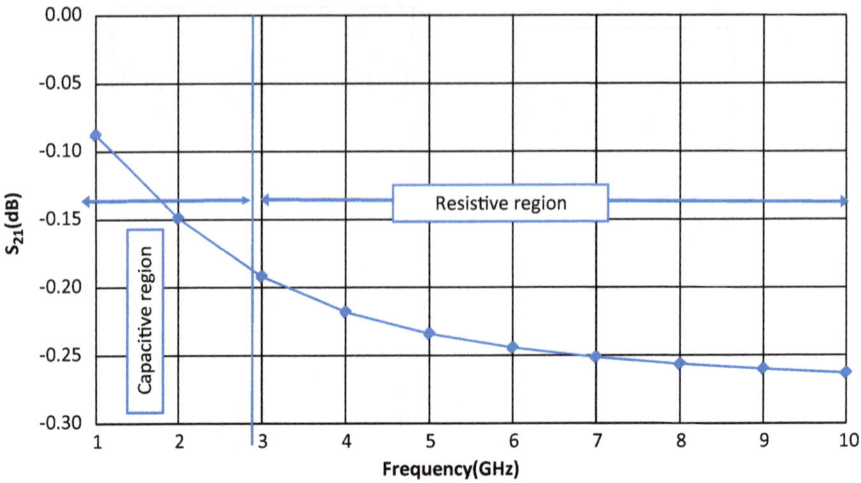

Fig. 3.11 Extracted S_{21} magnitude by using the quasi-static simulator

Scattering matrix (S) is a way of calculating a two-port network in terms of incident and reflected (or scattered) power. Scattering matrix is found by the following set of equations:

$$\begin{bmatrix} V_1^- \\ V_2^- \\ \\ V_N^- \end{bmatrix} = \begin{bmatrix} S_{11} & S_{12} & & S_{1N} \\ S_{21} & & & \\ & & & \\ S_{N1} & & & S_{NN} \end{bmatrix} \begin{bmatrix} V_1^+ \\ V_2^+ \\ \\ V_N^+ \end{bmatrix}$$

where,

$S_{ij} = V_i^- / V_j^+$ when $V_k^+ = 0$ for $k \neq j$.

1. Terminate all ports except j with matched load.
2. Drive port j with V_j^+
3. Measure reflected voltage V_i^- on port i.

In a two-port passive system, S_{11} equals S_{22}, and S_{12} equals S_{21}. Based on the quasi-static field solver simulation, the magnitude of S_{21} drops sharply for frequencies >2 GHz because the lossy characteristic of silicon starts to appear as the impedance of the oxide capacitance decreases as illustrated in Fig. 3.11.

From the EM simulations, it is observed that in all cases, the equivalent resistance and inductance of a TSV are affected by the arrangement of the TSVs around it, but this effect can be neglected (Fig. 3.12). The TSV coupling capacitance, inductance, and resistance increase as the spacing between the TSVs decreases. Moreover, the arrangement of the TSVs also affects the coupling capacitance and resistance while the coupling inductance almost remains constant (Fig. 3.13).

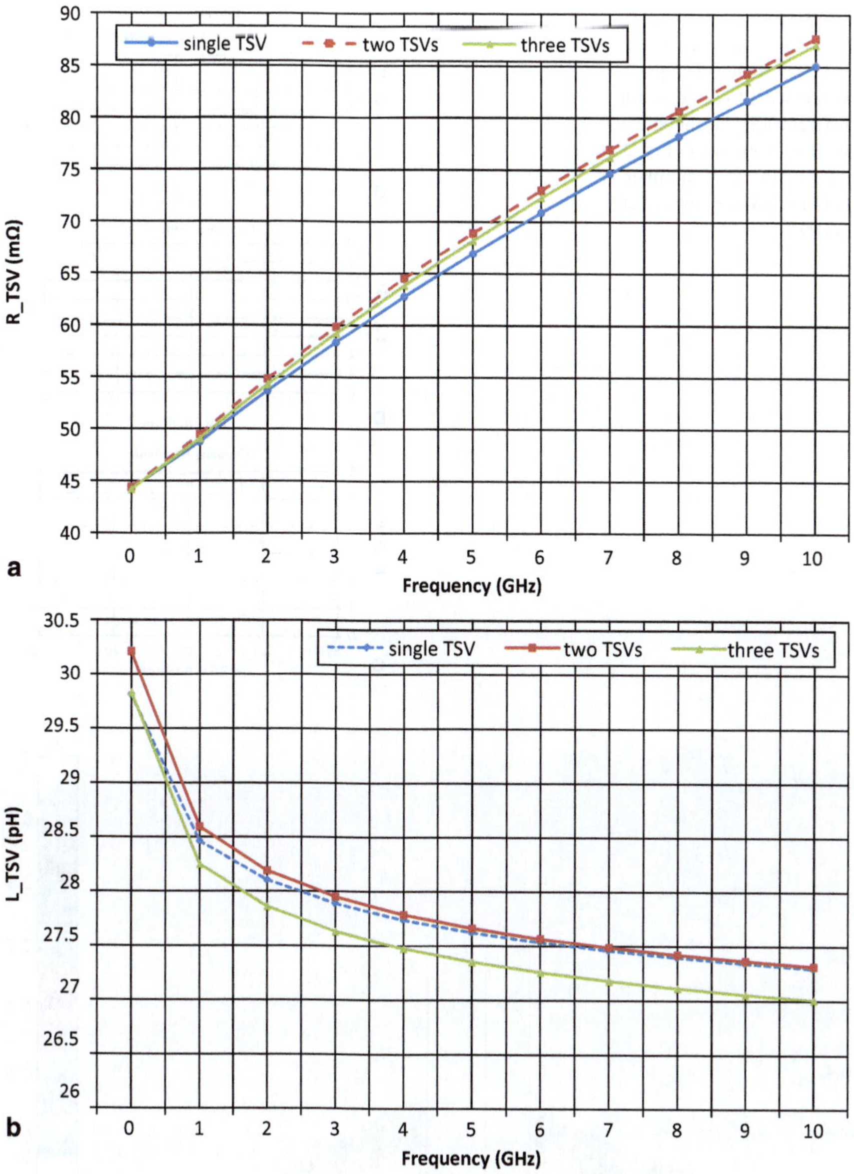

Fig. 3.12 The equivalent **a** self-resistance and **b** self-inductance of a TSV for three test cases: Single TSV, two straightly arranged TSVs, and three straightly arranged TSVs

Also, through simulating a 5 × 5 array of TSVs, obtaining the capacitance coupling coefficient matrix as an output, and plotting the row of the central TSV in this matrix, produces Fig. 3.14. It is shown that the capacitive coupling is almost insignificant beyond the nearest neighbors surrounding the TSV in the center of the

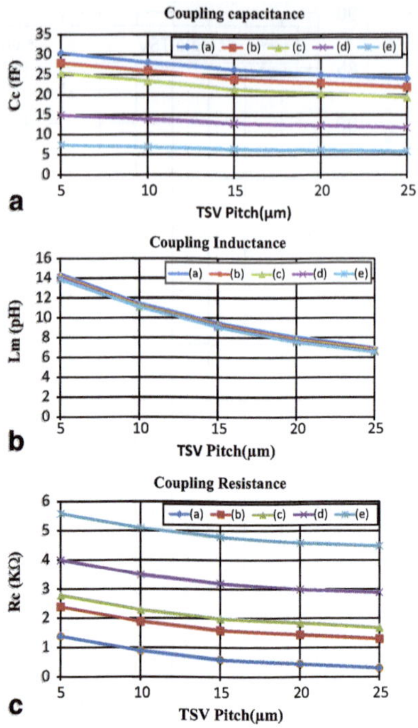

Fig. 3.13 The coupling **a** capacitance, **b** inductance, and **c** resistance change with respect to distance for different TSV arrangements previously shown in Fig. 3.10 between the TSV (number 1) and the TSV (number 2) at 1 GHz

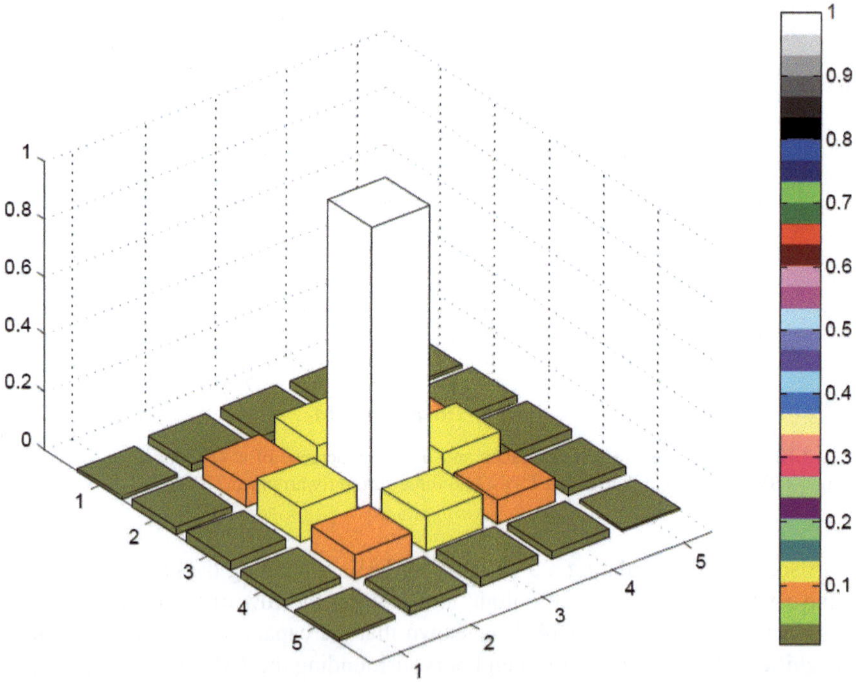

Fig. 3.14 Capacitive coupling normalized with respect to the total capacitance of the TSV in center

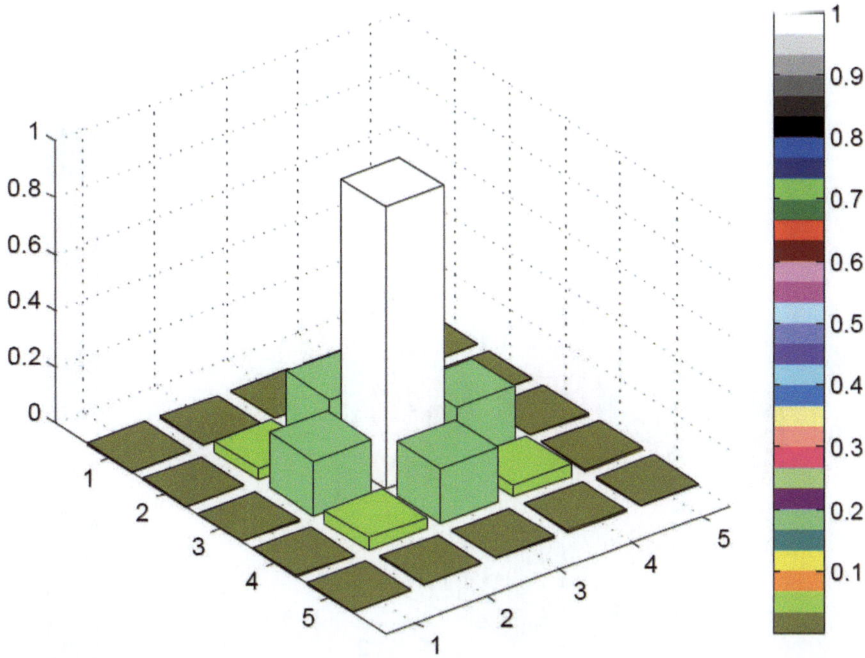

Fig. 3.15 Resistive coupling normalized with respect to the total capacitance of the TSV in the center

array. Also, coupling in case of the nearest adjacent TSVs with respect to the central TSV is higher than the coupling with the TSVs in the direction of the diagonal. In case of the resistive coupling, the same can be demonstrated through Fig. 3.15.

On the contrary of the short range of both capacitive and resistive coupling, inductive coupling has long range which extends to the second neighbors (Fig. 3.16). This is due to the fact that capacitive coupling is formed due the electric field lines while inductive coupling results from magnetic field lines. The electric field lines are almost totally shielded through the metals surrounding the central TSV; however, magnetic field lines are not.

Moreover, the center TSV has better performance than the corner TSV because in the case of center TSV the shielding effects of the surrounding TSVs will keep the energy from radiating out. Also, the cross talk between the corner TSVs is considerably stronger at high frequencies because of constructive interference of the scattered fields from the neighboring TSVs. Further, due to the larger separation of diagonal TSVs, the diagonal cross talk is usually weaker than the adjacent cross talk.

3.2.6.1 TSV Resistance Modeling

To model the resistance of the TSV (R_{TSV}) as a function of the physical dimensions, three variables come into play: the length of the TSV (l_{tsv}), the radius of the TSV

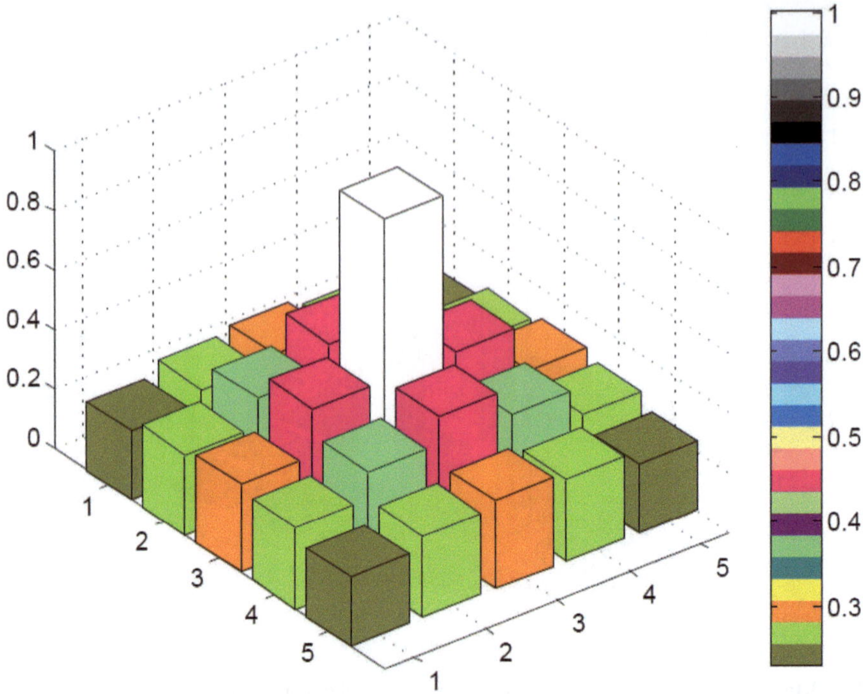

Fig. 3.16 Inductive coupling normalized with respect to the total inductance of the central TSV

(r_{tsv}), and the conductivity of the TSV material (σ); therefore, the resistance of the TSV can be generally expressed as

$$R_{TSV} = f(\sigma, l_{tsv}, r_{tsv}) \tag{3.10}$$

The dimensions are $\sigma\left[\Omega^{-1}m^{-1}\right]$, l_{tsv}[m], r_{tsv}[m], and R[Ω].

Therefore, Eq. (3.10) can be expressed in more details as

$$R_{TSV}[\Omega] = f\left(\left[\sigma\,\Omega^{-1}m^{-1}\right], l_{tsv}[m], r_{tsv}[m]\right) \tag{3.11}$$

Dimensional analysis can be used to simplify the analysis of this complicated problem. The dependencies between the variables will be established by using the π theorem. To start, construct a table for the relation between variables and dimensions; the number of dimensions is two and the number of variables is four. As a rule, the number of dimensions should always be less than the number of variables.

The dimensional matrix of the problem is given in Table 3.7. Then calculate certain matrix operations: [A], [B], $-[A^{-1}\cdot B]^T$ as depicted in Table 3.8, where [A] should not have a column that all are zero or same number and should be $n \times n$ matrix. If matrix A is not singular, interchange two columns (variables) in the

Table 3.7 The dimensional matrix of the R_{TSV} modeling problem

	R	l_{tsv}	σ	r_{tsv}
Ω	1	0	−1	0
m	0	1	−1	1

Table 3.8 Calculate D and C matrix of the R_{TSV} modeling problem

	R	l_{tsv}	σ	r_{tsv}
Ω	1	0 [B]	-1	0 [A]
m	0	1	-1	1
π_1	[I]		$-[A^{-1}.B]^T$	
π_2				

Table 3.9 π terms calculations for the R_{TSV} modeling problem

	R	l_{tsv}	σ	r_{tsv}
Ω	1	0	-1	0
m	0	1	-1	1
π_1	1	0	1	1
π_2	0	1	0	-1

dimensional matrix to make matrix A nonsingular. Subsequently, the π terms of this problem are given by Table 3.9:

$$\pi_1 = R_{TSV}\sigma r_{tsv} \tag{3.12}$$

$$\pi_2 = \frac{l_{tsv}}{r_{tsv}} \tag{3.13}$$

From the π theorem, we can relate the variables to one another via a function, where

$$\pi_1 = f(\pi_2) \tag{3.14}$$

Substituting (3.12) and (3.13) into (3.14) results in

$$R_{TSV}\sigma r_{tsv} = f\left(\frac{l_{tsv}}{r_{tsv}}\right) \tag{3.15}$$

Now the function of TSV resistance is related to one variable $\left(\frac{l_{tsv}}{r_{tsv}}\right)$ instead of three variables $(\sigma, l_{tsv}, r_{tsv})$. Finally, the function is determined by using a curve fitting technique.

Another method is described here to model the resistance using the dimensional analysis method. By recalling Eq. (3.11):

$$R_{TSV}[\Omega] = f\left(\sigma\left[\Omega^{-1}m^{-1}\right], l_{tsv}[m], r_{tsv}[m]\right) \tag{3.16}$$

Propose

$$R_{TSV}\Omega = \sigma^a l_{tsv}{}^b r_{tsv}{}^c \tag{3.17}$$

Equating the powers of the dimensions results in

$$\Omega = \Omega^{-a}\ \mathrm{m}^{-a}\ \mathrm{m}^{b}\ \mathrm{m}^{c} \tag{3.18}$$

$$\Omega : 1 = -a \tag{3.19}$$

$$m : b + c - a = 0 \tag{3.20}$$

From (3.19) and (3.20):

$$a = -1 \tag{3.21}$$

$$b + c = -1 \tag{3.22}$$

Let $b = 1$, and substituting in (3.22) results in

$$c = -2 \tag{3.23}$$

Substituting (3.13), (3.14), and (3.15) into (3.9) results in

$$R_{TSV} = \alpha\sigma^{-1} l_{tsv} r_{tsv}{}^{-2} \tag{3.24}$$

where α is a constant and can be found by a curve fitting technique.

3.2.6.2 TSV Capacitance Modeling

To model the capacitance of the TSV (C_{TSV}) as a function of the physical dimensions, four variables come into play: the length of the TSV (l_{tsv}), the radius of the TSV (r_{tsv}), the electric permittivity of the dielectric material surrounding the TSV (ε), and the oxide thickness (t_{ox}). Therefore, the capacitance of the TSV can be generally expressed as

Table 3.10 The dimensional matrix of the C_{TSV} modeling problem

	C	l_{tsv}	ε	r_{tsv}	t_{ox}
Ω	-1	-1 [B]	0	0	0 [A]
m	0	-1	1	1	1
S	1	1	0	0	0
π_1	[I]		$-[A^{-1}.B]^T$		
π_2					

Table 3.11 The dimensional matrix of the C_{TSV} modeling problem after rearranging

	C	ε	l_{tsv}	r_{tsv}	t_{ox}
Ω	-1	0	-1	0	0
m	0	1	-1	1	1
S	1	0	1	0	0
π_1	[I]		$-[A^{-1}.B]^T$		
π_2					

$$C_{TSV}[\Omega^{-1}s] = f(\varepsilon[\Omega^{-1}m^{-1}s], l_{tsv}[m], \mathrm{r_{tsv}}[m], \mathrm{t_{ox}}[m]) \tag{3.25}$$

Assume we arranged the variables as in Table 3.10; this arrangement is wrong because there is a complete line equal to zero in [A], so rearrange them as in Table 3.11. The third row is linearly dependent on the first row, i.e., the third row is repeated with the first one, so remove it. Now, $[A]=2\times 2$ and $[B]=2\times 3$, the matrix rank is 2, and the number of π terms is the order of matrix B which is 3; the π terms of this problem are given by Table 3.12:

$$\pi_1 = \frac{C_{TSV}}{\varepsilon l_{tsv}} \tag{3.26}$$

$$\pi_2 = \frac{t_{ox}}{l_{tsv}} \tag{3.27}$$

$$\pi_3 = \frac{r_{tsv}}{l_{tsv}} \tag{3.28}$$

Table 3.12 π terms calculations for the C_{TSV} modeling problem

	C	l_{tsv}	ε	r_{tsv}	t_{ox}
Ω	-1	0	0	-1	0
m	0	1	1	-1	1
π_1	1	0	0	-1	-1
π_2	0	1	0	0	-1
π_3	0	0	1	0	-1

$$\pi_1 = f(\pi_2, \pi_3) \tag{3.29}$$

Substituting (3.26), (3.27), and (3.28) into (3.29) results in

$$\frac{C_{TSV}}{\varepsilon\, l_{tsv}} = f\left(\frac{t_{ox}}{l_{tsv}}, \frac{r_{tsv}}{l_{tsv}}\right) \tag{3.30}$$

Now the function of TSV capacitance is related to two variables ($\frac{t_{ox}}{l_{tsv}}, \frac{r_{tsv}}{l_{tsv}}$) instead of four variables ($\varepsilon, l_{tsv}, r_{tsv}, t_{ox}$). Using the curve fitting technique, the TSV capacitance is given by

$$C_{TSV} = \frac{2.7\pi\varepsilon_0\varepsilon_{ox} l_{tsv}}{ln\left(1 + \frac{r_{tsv} + t_{ox}}{l_{tsv}}\right)} \tag{3.31}$$

3.2.6.3 TSV Inductance Modeling

To model the inductance of the TSV (L_{TSV}) as a function of the physical dimensions, three variables come into play: the length of the TSV (l_{tsv}), the radius of the TSV (r_{tsv}), and the magnetic permeability of the material surrounding the TSV (μ). Therefore, the inductance of the TSV can be generally expressed as

$$L_{TSV}\left[\Omega s\right] = f(\mu[\Omega\, m^{-1} s], l_{tsv}[m], r_{tsv}[m]) \tag{3.32}$$

The π terms of this problem are given by Table 3.13:

$$\pi_1 = \frac{L_{TSV}}{\mu l_{tsv}} \tag{3.33}$$

$$\pi_2 = \frac{r_{tsv}}{l_{tsv}} \tag{3.34}$$

Table 3.13 π terms calculations for the L_{TSV} modeling problem

	I	r_{tsv}	μ	l_{tsv}
Ω	1	0	1	0
m	0	1	-1	1
S	1	0	1	0
π_1	1	0	-1	-1
π_2	0	1	0	-1

$$\pi_1 = f(\pi_2) \tag{3.35}$$

Substituting (3.33) and (3.34) into (3.35) results in

$$\frac{L_{TSV}}{\mu l_{tsv}} = f\pi\left(\frac{r_{tsv}}{l_{tsv}}\right) \tag{3.36}$$

Now the function of TSV inductance is related to one variable ($\frac{r_{tsv}}{l_{tsv}}$) instead of three variables (μ, l_{tsv}, r_{tsv}). Using the curve fitting technique, the TSV inductance is given by

$$L_{TSV} = \frac{170\,\mu_0 L_{tsv}}{2\pi} ln\left(1 + 0.01\frac{l_{tsv}}{r_{tsv}}\right) \tag{3.37}$$

3.2.6.4 Final Element Formulae Using Curve Fitting

Applying the dimensional analysis to the capacitance of the TSV results in

$$\frac{C}{\varepsilon_{ox} r_{tsv}} = f\left(\frac{l_{tsv}}{r_{tsv}}, \frac{t_{ox}}{r_{tsv}}\right) \tag{3.38}$$

Any capacitance in the model depends on only two dimensionless variables instead of four physical variables. Similarly, resistance and inductance have the following dependency:

$$\mathrm{R}\,\sigma\,\mathrm{r}_{tsv} = \mathrm{f}\left(\frac{l_{tsv}}{\mathrm{r}_{tsv}}\right) \tag{3.39}$$

$$\frac{\mathrm{L}}{\mu\,\mathrm{r}_{tsv}} = \mathrm{f}\left(\frac{l_{tsv}}{\mathrm{r}_{tsv}}\right) \tag{3.40}$$

The formulas for the self and coupling capacitance, resistance, and inductance of the TSV were found to be approximated well by the following set of equations:

$$L_0 = \frac{155\,\mu_0 r_{tsv}}{2\pi} \ln\left(1 + 0.01\frac{l_{tsv}}{r_{tsv}}\right) \tag{3.41}$$

$$L_1 = \frac{55\,\mu_0 r_{tsv}}{2\pi} \ln\left(1 + 0.01\frac{l_{tsv}}{r_{tsv}}\right) \tag{3.42}$$

$$R_0 = \frac{330 \ln\left(1 + 0.01\frac{l_{tsv}}{r_{tsv}}\right)}{2\pi\,\sigma_c r_{tsv}} \tag{3.43}$$

$$R_1 = \frac{200 \ln\left(1 + 0.01\frac{l_{tsv}}{r_{tsv}}\right)}{2\pi\,\sigma_c r_{tsv}} \tag{3.44}$$

$$C_{ox} = \frac{3.3\pi\varepsilon_0\varepsilon_{0x} l_{tsv}}{\ln\left(1 + \frac{t_{ox}}{r_{tsv}}\right)} \tag{3.45}$$

$$C_{si} = \frac{0.5\,\pi\left(\frac{p_{tsv}}{r_{tsv}}\right)^{\alpha_1}(8n_{bc} + 1)\varepsilon_0\varepsilon_{si} l_{tsv}}{\ln\left(1 + \frac{w_{bc}}{r_{tsv}}\right)} \tag{3.46}$$

$$C_{si} = \frac{\left(\frac{p_{tsv}}{r_{tsv}}\right)^{\alpha_2} \ln\left(1 + \frac{w_{bc}}{r_{tsv}}\right)}{0.5\,\pi(8n_{bc} + 1)\sigma_{si} l_{tsv}} \tag{3.47}$$

where α_1, α_2 are determined according to the test case, and their corresponding values are shown in Table 3.14. From Eqs. (3.45) and (3.46), it is observed that the value of the silicon capacitance is highly dependent on the body contacts around the TSV. In the extreme case when there are no body contacts, this capacitance becomes similar to the stray capacitance, which is small (in the order of several fF).

Table 3.14 Values of α_1, α_2, α_3, α_4 and their corresponding test cases

Test case	α_1	α_2	α_3	α_4
Single TSVs	0	0	0	0
Two TSVs	1.6	1.2	1	1
Three TSVs	1.8	1.3	1.6	1.5

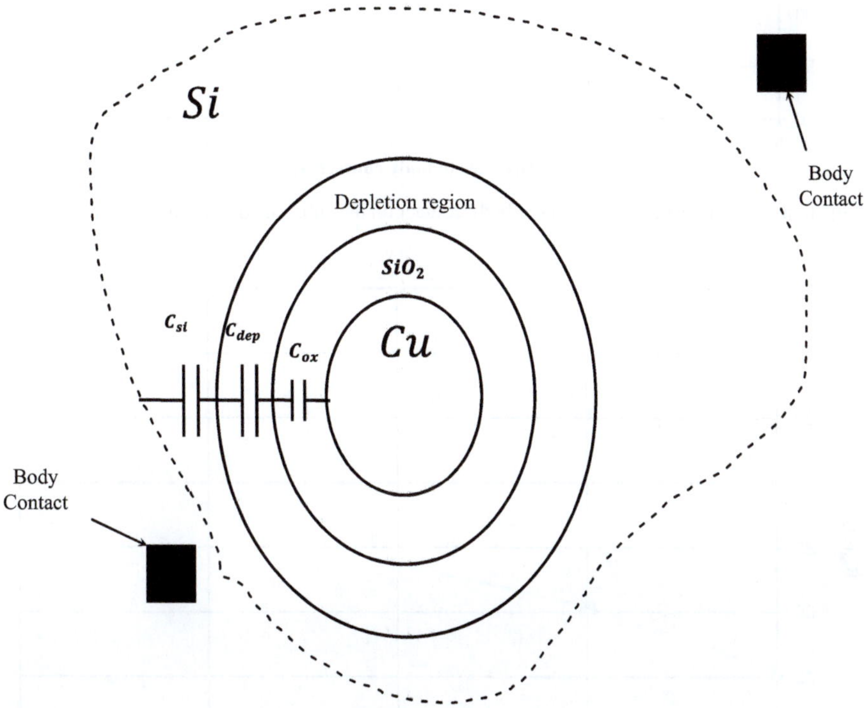

Fig 3.17 Top view of a single TSV, including a virtual ground, due to body contacts

However, the presence of the body contacts around the TSV creates a virtual ground surface around the TSV (Fig. 3.17). In this case, the silicon capacitance increases to several tens of fF. The closer the body contacts are to a TSV and the denser they are, the more the TSV capacitance and the isolation of the TSV from other circuit components.

The impact of distance of the body contact at $n_{bc} = 4$ on the substrate capacitance and resistance is shown in Figs. 3.18 and 3.19, respectively, for a single TSV.

The depletion capacitance and resistance are given by

$$C_{dep} = \frac{2\pi\varepsilon_0\varepsilon_{si} l_{tsv}}{\ln\left(1+\frac{t_{ox}+w_{dep}}{r_{tsv}}\right)\sqrt{1+\frac{V_{tsv}}{V_{th}}}} \tag{3.48}$$

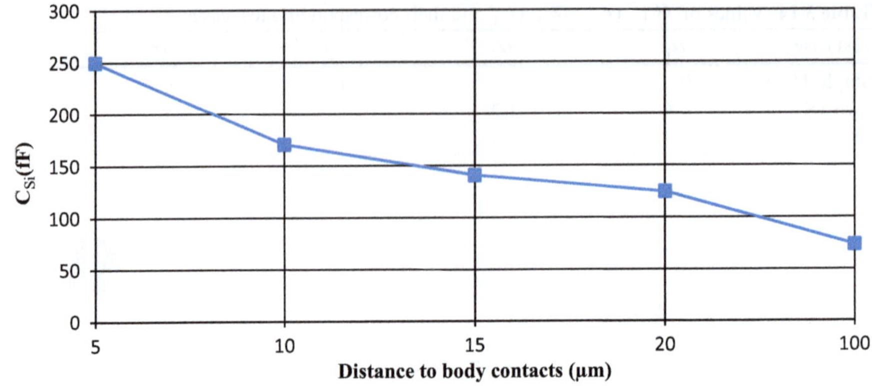

Fig. 3.18 Impact of the distance of the body contact on the substrate depletion capacitance

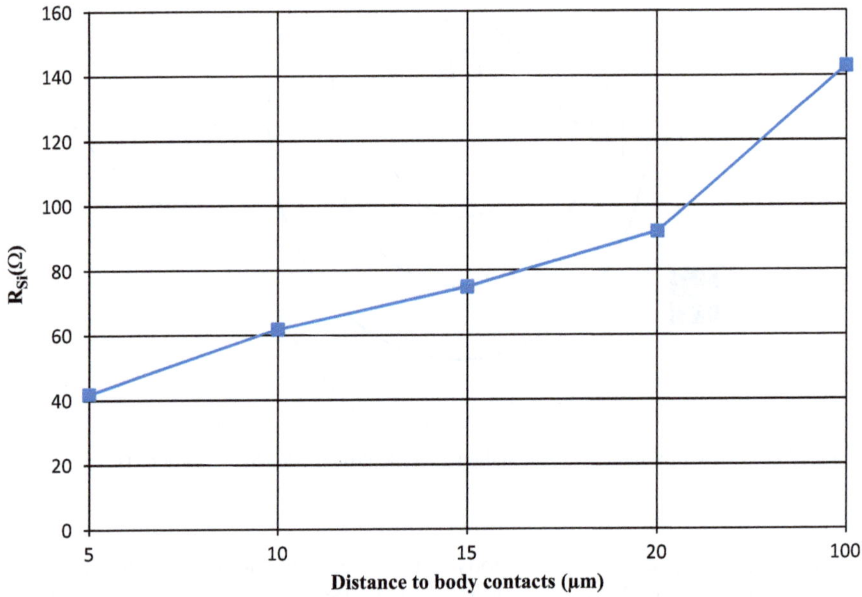

Fig. 3.19 Impact of the distance of the body contact on the substrate depletion resistance

$$R_{dep} = \frac{\ln\left(1 + \frac{t_{ox} + w_{dep}}{r_{tsv}}\right)\sqrt{1 + \frac{V_{tsv}}{V_{th}}}}{2\pi\sigma_{si} l_{tsv}} \tag{3.49}$$

To realize a voltage-dependent depletion capacitance and resistance in a SPICE-compatible format, the first-order Taylor expansion can be used to replace the

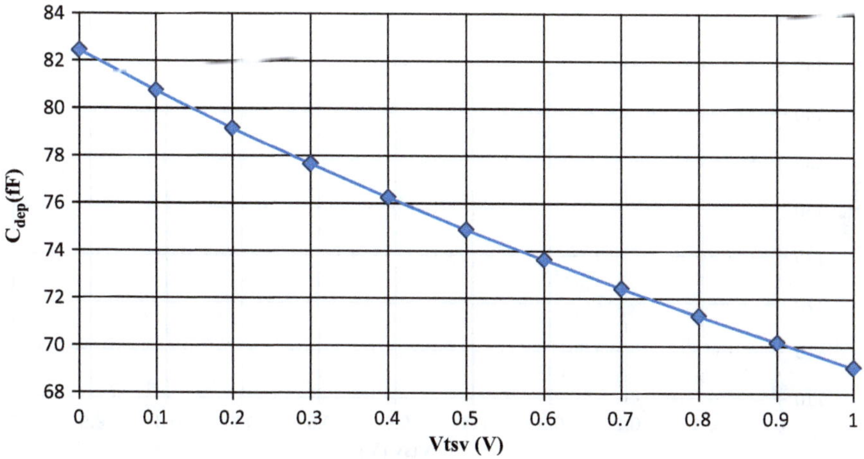

Fig. 3.20 Depletion capacitance change with respect to the voltage applied on the TSV (p-type Si)

$\sqrt{1+\frac{V_{tsv}}{V_{th}}}$ term in (3.48) and (3.49), where $V_{tsv} < V_{th}$, for typical geometrical and electrical parameters of the TSV ($V_{th} = 2.1\text{V}$), and for the voltage range ($0\text{V} \le V_{tsv} \le 1\text{V}$):

$$\frac{1}{\sqrt{1+\frac{V_{tsv}}{V_{th}}}} = 1-\frac{V_{tsv}}{2V_{th}} \tag{3.50}$$

$$\sqrt{1+\frac{V_{tsv}}{V_{th}}} = 1+\frac{V_{tsv}}{2V_{th}} \tag{3.51}$$

Substituting (3.50) into (3.48) and (3.51) into (3.49) results in

$$C_{dep} = \frac{2\pi\varepsilon_0\varepsilon_{si}l_{tsv}}{\ln\left(1+\frac{t_{ox}+w_{dep}}{r_{tsv}}\right)}\left(1-\frac{V_{tsv}}{2V_{th}}\right) = C_{dep0} - \beta_1 V_{tsv} \tag{3.52}$$

$$R_{dep} = \frac{\ln\left(1+\frac{t_{ox}+w_{dep}}{r_{tsv}}\right)}{2\pi\sigma_{si}l_{tsv}}\left(1+\frac{V_{tsv}}{2V_{th}}\right) = R_{dep0} + \beta_2 V_{tsv} \tag{3.53}$$

Most SPICE-like dynamic simulators accept (3.52) and (3.53) at entry level.

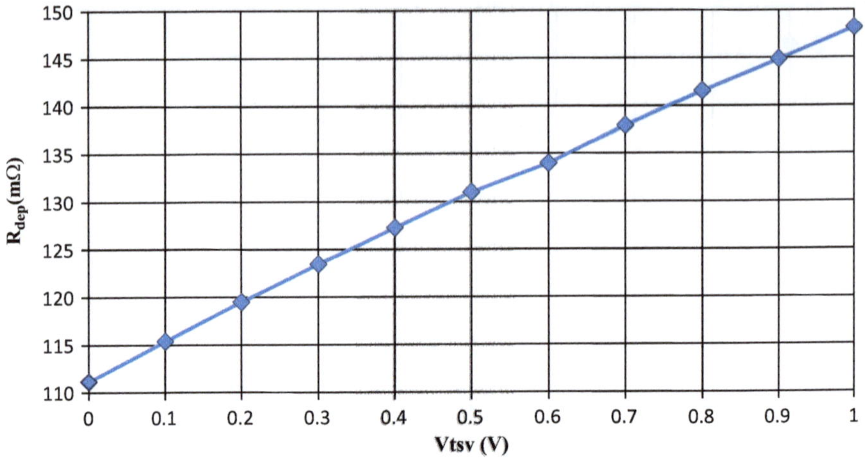

Fig. 3.21 Depletion resistance change with respect to the voltage applied on the TSV (p-type Si)

The depletion capacitance and resistance are highly nonlinear and change by more than 100% in the range between 0 and 1 V. The change of the depletion capacitance and resistance with respect to the voltage applied on the TSV is shown in Figs. 3.20 and 3.21, respectively. Also, none of the previous related publications have obtained a formula for the depletion resistance.

As stated earlier, from the EM simulations, it is shown that the capacitive and resistive coupling is almost insignificant beyond the first neighbors surrounding the TSV while inductive coupling has a long range extending to far neighbors.

The coupling capacitance and resistance between the TSVs number 1 and 2 in Fig. 3.10 are given by

$$C_{c1} = \frac{1.6\pi\varepsilon_0\varepsilon_{si}l_{tsv}}{\alpha_3 ln\left(1+\frac{p_{tsv}}{r_{tsv}}\right)} \tag{3.54}$$

$$R_{c1} = \frac{\alpha_4 ln\left(1+\frac{p_{tsv}}{r_{tsv}}\right)}{0.75\pi\sigma_{si}l_{tsv}} \tag{3.55}$$

where α_3 and α_4 are determined according to the test case, and their corresponding values are shown in Table 3.14. The mutual inductance is given by

$$L_{mi} = K_i L_m \tag{3.56}$$

where L_m is given by

$$L_m = \frac{434\mu_0 r_{tsv}}{2\pi} ln\left(1 + 0.02\frac{l_{tsv}}{P_{tsv}}\right), \quad (3.57)$$

and the coupling coefficient (K_i) is given by

$$K_i = \frac{1}{2i}, \quad (3.58)$$

where $i = 1$ for the first neighbor and $i = 2$ for the second neighbor and so on. These closed-form expressions represent accurate and direct parasitics extraction of multiple-TSV structures for any technology parameter set and TSV geometry as well as body contacts' distance and number.

Typical extracted values of the lumped element model components for single- and multiple-TSV structures are given in Table 3.15.

Typical values of different interconnection technologies parasitic parameters as compared to TSV parasitics are summarized in Table 3.16 [14]. It is clear that the TSV technology provides the best performance.

3.3 TSV Model Linearization

In Sect. 3.2.5, we obtained closed-form expressions for the depletion resistance and capacitance. In this section, we will drive a SPICE-compatible model for the nonlinear capacitance and resistance as follows: The voltage-dependent capacitance (left side of Fig. 3.22) is given by

$$C_t = C_0 + \alpha V \quad (3.59)$$

The current across it is given by

$$I = j\omega(C_0 + \alpha V)V = j\omega C_0 V + j\omega\alpha V^2 \quad (3.60)$$

For the equivalent SPICE-compatible model (right side of Fig. 3.22), the current into the capacitor is given by

$$I = j\omega C_0(V - V_{vcvc}) = j\omega C_0 V - j\omega C_0 V_{vcvc} \quad (3.61)$$

From (3.69) and (3.70):

$$V_{vcvc} = -\frac{\alpha}{C_0}V^2 \quad (3.62)$$

Table 3.15 Extracted values for *R*, *L*, and C of multiple-TSV structures from the closed-form formulas

Test case	$R(\Omega)$		$C(fF)$		$L(pH)$	
Single TSV (n_{bc}=0)	R_0	63 m	C_{ox}	124	L_0	14
	R_1	38 m	C_{si}	3	L_1	5
	R_{si}	1.2 K	C_{dep}	78		
	R_{dep}	124				
Single TSV (n_{bc} =4)	R_0	63 m	C_{ox}	124	L_0	14
	R_1	38 m	C_{si}	250	L_1	5
	R_{si}	55	C_{dep}	78		
	R_{dep}	124				
Two TSVs (n_{bc} =0)	R_0	63 m	C_{ox}	124	L_0	14
	R_1	38 m	C_{si}	4.6	L_1	5
	R_{si}	1.4 K	C_{dep}	78	L_{m1}	14
	R_{dep}	124	C_{c1}	13		
	R_{c1}	3.36 K				
Three TSVs (n_{bc} =0)	R_0	63 m	C_{ox}	124	L_0	14
	R_1	38 m	C_{si}	5.3	L_1	5
	R_{si}	1.8 K	C_{dep}	78	L_{m1}	14
	R_{dep}	124	C_{c1}	8	L_{m2}	7
	R_{c1}	5 K				

Table 3.16 Typical values of parasitic components for different chip interconnection technologies

Technology	TSV	Wire bond	Solder bump	Adhesive bumps
Resistance (mΩ)	10–60	30–100	1–3	15–30
Inductance (pH)	5–10	1000–3000	50–100	50–100
Capacitance (fF)	3–10	10–50	2–10	2–10

The coefficients of the voltage-controlled voltage source can be expressed as

$$V_{vcvc} = \left(0, 0, -\frac{\alpha}{C_0}\right) \tag{3.63}$$

Hence, the voltage-dependent capacitor is modeled using a constant capacitor and a voltage-controlled voltage source (V_{vcvc}) [15].

In a similar way, the voltage-dependent resistance (left side of Fig. 3.23) is given by

$$R_t = R_0 + \alpha V \tag{3.64}$$

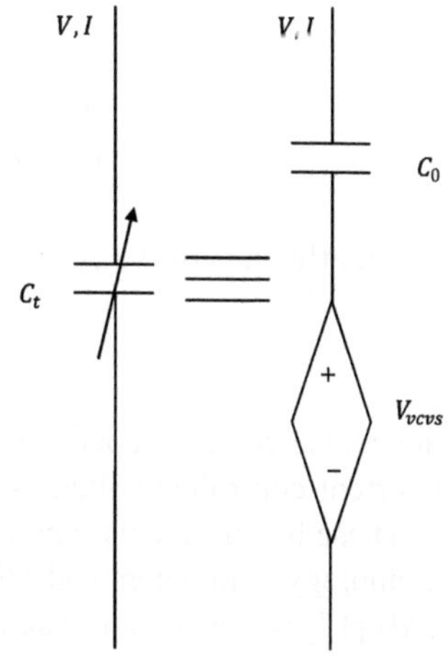

Fig. 3.22 The equivalent circuit model of the voltage-dependent capacitance

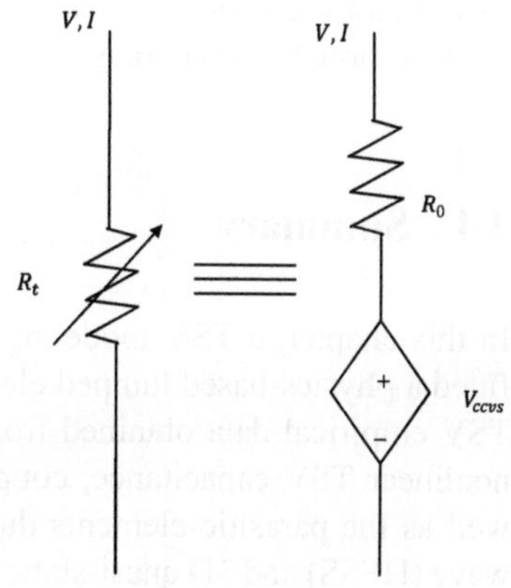

Fig. 3.23 The equivalent circuit model of the voltage-dependent resistance

The voltage across it is given by

$$V = IR_0 + I\alpha V \tag{3.65}$$

For the equivalent SPICE-compatible model (right side of Fig. 3.23), the voltage across it is given by

$$V = IR_0 + V_{ccvc} \tag{3.66}$$

From (3.37) and (3.38):

$$V_{ccvc} = I\alpha V = I\alpha(IR_0 + V_{ccvc}) = \alpha R_0 I^2 + I\,\alpha V_{ccvc} \tag{3.67}$$

$$(I - I\alpha)V_{ccvc} = \alpha R_0 I^2 \tag{3.68}$$

First-order Taylor expansion is applied to (3.40) which results in [16]

$$V_{ccvc} = \frac{\alpha R_0 I^2}{(1 - I\alpha)} = \alpha R_0 I^2 (1 + I\alpha) = \alpha R_0 I^2 + \alpha^2 R_0 I^3 \tag{3.69}$$

The coefficients of the current-controlled voltage source are expressed as

$$V_{ccvc} = (0, 0, \alpha R_0, \alpha^2 R_0) \tag{3.70}$$

Hence, the voltage-dependent resistance is modeled using a constant resistance and a current-controlled voltage source (V_{ccvc}).

These linear models represent accurate and direct extraction of the TSV for any technology parameters and TSV geometry for frequency range: 0–100 GHz. MATLAB [17] is used to simulate the exact nonlinear resistance and capacitance model, while SPICE is used to simulate the equivalent model.

A voltage pulse source, with a 100 Ω internal resistance, is applied to the test model and the results are shown in Figs. 3.24 and 3.25, respectively, where the models show high accuracy.

3.4 Summary

In this chapter, a TSV modeling methodology was introduced. The methodology fitted a physics-based lumped element model to the complete frequency-dependent TSV empirical data obtained from a field solver. The model includes skin effect, nonlinear TSV capacitance, coupling parasitics, adjacent body contact effects, as well as the parasitic elements due to the finite substrate resistivity. Both 3D full-wave (HFSS) and 3D quasi-static (Q3D) field solvers were considered. Comparison between those two field solvers from few Hertz to 100 GHz showed that the difference in accuracy is <5 %; hence, the Q3D field solver was used. Dimensional analysis was used to develop closed-form formulas for the lumped elements. This chapter also showed that the capacitance of the TSV is highly dependent on the body contact placement and count. Moreover, we represented the TSV nonlinear elements with multiple linear elements.

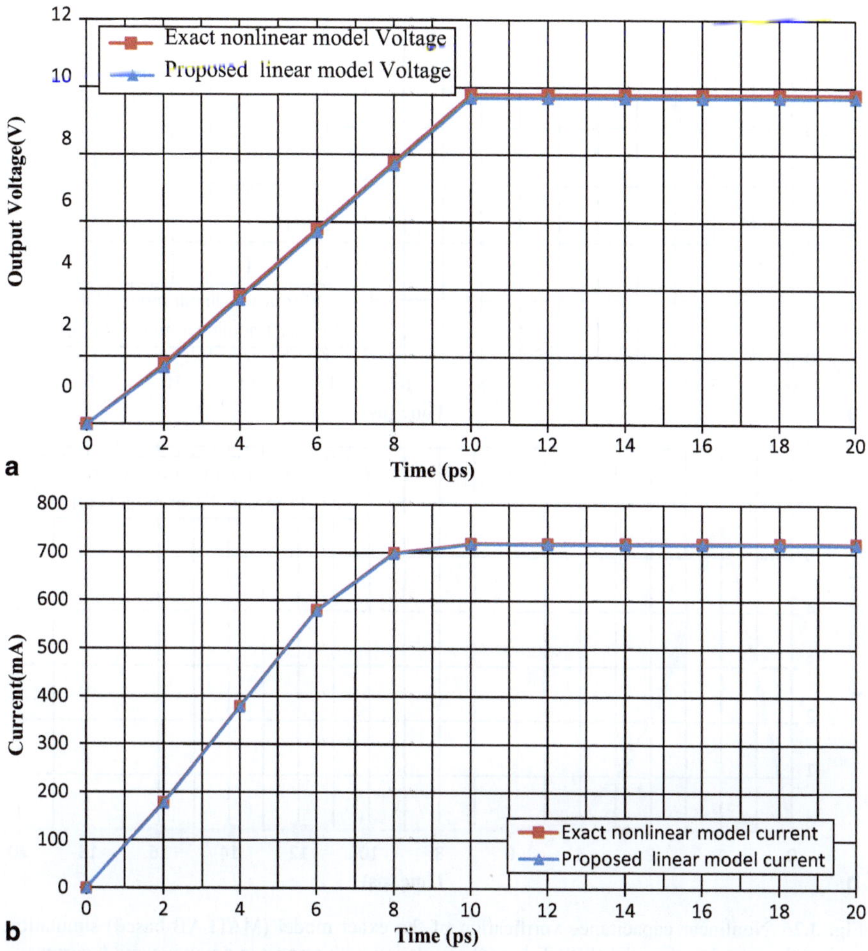

Fig. 3.24 Nonlinear resistance verification of the exact model (MATLAB-based) simulation against the model (SPICE-based) simulation with respect to **a** voltage and **b** current

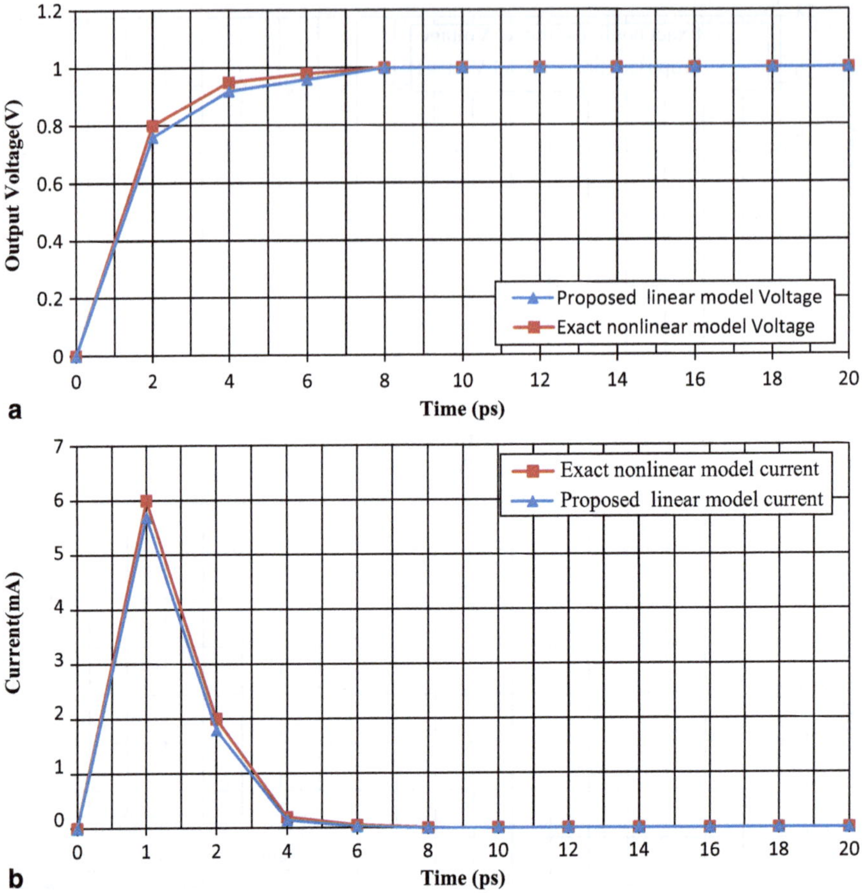

Fig. 3.25 Nonlinear capacitance verification of the exact model (MATLAB-based) simulation against the introduced model (SPICE-based) simulation with respect to **a** voltage and **b** current

References

1. http://www.itrs.net
2. L. Cadix, C. Fuchs, M. Rousseau, P. Leduc, H. Chaabouni, A. Thuaire, M. Brocard, A. Valentian, A. Farcy, C. Bermond, N. Sillon, B. Fléchet, P. Ancey, Integration and frequency dependent parametric modeling of through silicon via involved in high density 3D chip stacking
3. http://www.eda.ncsu.edu/wiki/freePDK
4. R. Weerasekera, L.-R. Zheng, D. Pamunuwa, H. Tenhunen, *Extending systems-on-chip to the third dimension: performance, cost and technological trade-off*, in Proceedings of IEEE/ACM International Conference on Computer-Aided Design (ICCAD), 2007, pp. 212–219
5. T. Szirtes, *Applied Dimensional Analysis and Modeling*. (McGraw-Hill, New York, 1997)
6. K.J. Han, *Electromagnetic Modeling of Interconnections in Three-Dimensional Integration*. PhD Dissertation, Georgia Institute of Technology, Aug 2009

7. Ankur et al., New Paradigm in IC-Package Interconnections by Reworkable Nano-Interconnects, 2004 Electronic Components and Technology Conference, 2004, pp. 451–460
8. T. Bandyopadhyay, R. Chatterjee, D. Chung, M. Swaminathan, R. Tummala, *Electrical Modeling of Through Silicon and Package Vias*, in 3D System Integration. 3DIC 2009. IEEE International Conference, 2009
9. C. Xu, H. Li, R. Suaya, Compact AC Modeling and Analysis of Cu, W, and CNT based Through-Silicon Vias (TSVs) in 3-D ICs, in IEDM09-521, 2009
10. M. Stucchi, K.D. Meyer, W. Dehaene, S.G. Katti, *Electrical Modeling and Characterization of Through Silicon via for Three-Dimensional ICs*, IEEE Transactions on Electron Devices, vol. 57, no. 1, Jan 2010
11. Y.L. Chow, M.M.A. Salama, *A Simplified Method for Calculating the Substation Grounding Grid Resistance*, Power Delivery, IEEE Transactions,April 1994, pp. 736–742
12. http://www.ansoft.com/products/si/hfss
13. http://www.ansoft.com/products/si/q3d extractor
14. R. Fillion, C. Woychik, T. Zhang, D. Bitting, Embedded chip build-up using fine line interconnect, in Proceeding of the 57th Electronics Components and Technology Conference, 2007
15. A. Ito, A Voltage Dependent Capacitance Model Including Effects of Manufacturing Process Variabilities on Voltage Coefficients, IEEE Transactions on Computer-Aided Design of Integrated Circuits and Systems, vol. 14, no. 9, Sept 1995
16. http://en.wikipedia.org/wiki/Taylor_series
17. http://www.matlab.com

Chapter 4
TSV Verification

This chapter discusses the validation of the developed through silicon via (TSV) lumped element model and the closed form expressions of its elements' values versus the electromagnetic (EM) simulations in the frequency domain as well as the accuracy of the TSV model in the time domain and the analysis of the impact of a TSV on circuit performance. Moreover, the validation against device simulation is performed. The parameters' values of the TSV model are fitted to the simulated scattering (*S*) parameters up to 100 GHz with an error $<5\,\%$.

The robustness of the model enables direct extraction of the TSV resistance, self-inductance, oxide capacitance, as well as the parasitic elements due to the finite substrate resistivity. The model is compact and compatible with simulation program with integrated circuit emphasis (SPICE) simulators; hence, it allows fast investigation of the TSV impact on 3D circuits' performance.

Using the developed model, it is shown that a TSV has a negligible effect on high-impedance driver characteristics, and in such cases, the TSV can be modeled as a capacitance-dominated structure. In case of low-impedance devices, a TSV has a significant effect on the device characteristics and the full TSV model needs to be used. The model is also examined against device simulations, where we found a very good correlation with an error $<2\,\%$ as compared to full electromagnetic simulations of a TSV.

In order to fabricate 3D integrated circuits (ICs), new capabilities are needed. 3D process technologies are one of the most important topics that need to be investigated.

3D IC technologies can be broadly classified in two major categories: 3D package level technology (wire-bond-based) and 3D wafer-level technology (TSV-based). TSVs can be processed and fabricated before, during, or after the IC fabrication process, either for chip stacking or solely as an interposer.

K. Salah et al., *Arbitrary Modeling of TSVs for 3D Integrated Circuits*,
Analog Circuits and Signal Processing, DOI 10.1007/978-3-319-07611-9_4,

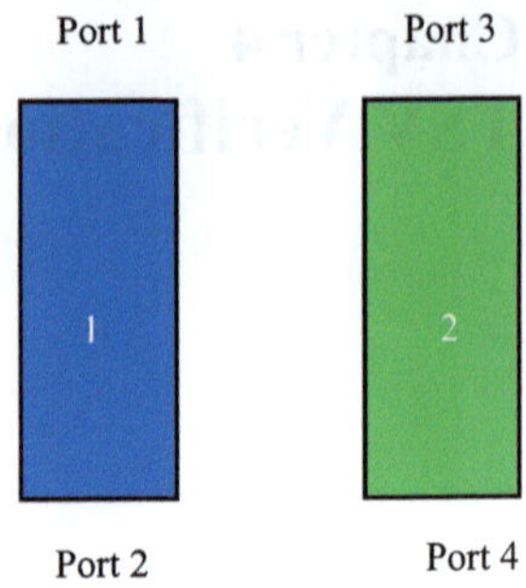

Fig. 4.1 Cross-section view of two straightly arranged TSVs and their ports, where the first TSV ports are 1, 2 and the second TSV ports are 3, 4

4.1 Validation Against Electromagnetic Simulation: Frequency Domain and Time Domain

The closed-form equations of the through silicon via (TSV) model elements are examined versus electromagnetic simulations for a single TSV and two linearly arranged TSVs (where the first TSV ports are 1, 2 and the second TSV ports are 3, 4 (Fig. 4.1), respectively. This comparison is shown in Figs. 4.2 and 4.3, where it shows a good correlation and an error <5 % for S_{21} and S_{31}.

To evaluate the time-domain accuracy of the TSV model, electromagnetic data from Q3D simulator [1] are used to produce a time-domain waveform using Agilent's Advanced Design System (ADS) environment [2]. This simulation is compared

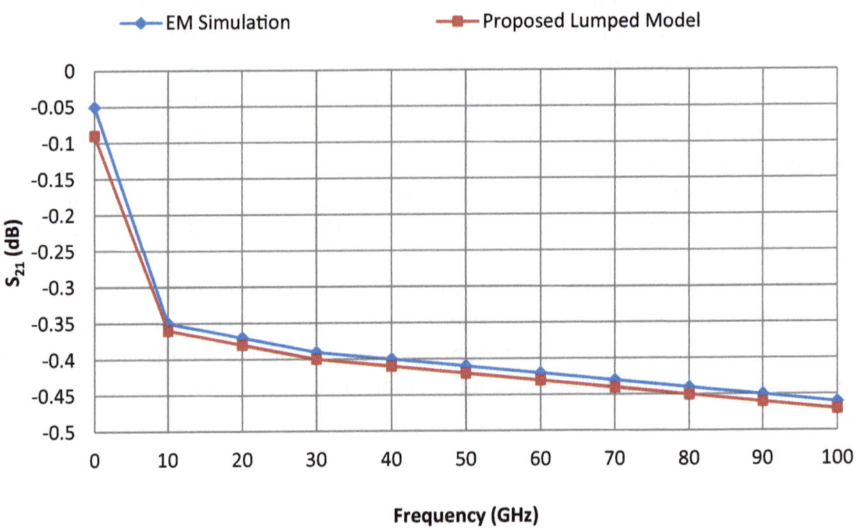

Fig. 4.2 A comparison of the quasi-static EM simulations against the lumped element model simulations for a single TSV ($r_{tsv} = 5\mu m$, $l_{tsv} = 50\mu m$, $V_{tsv} = 0.3Vm$, $\omega_{bc} = 5\mu m$)

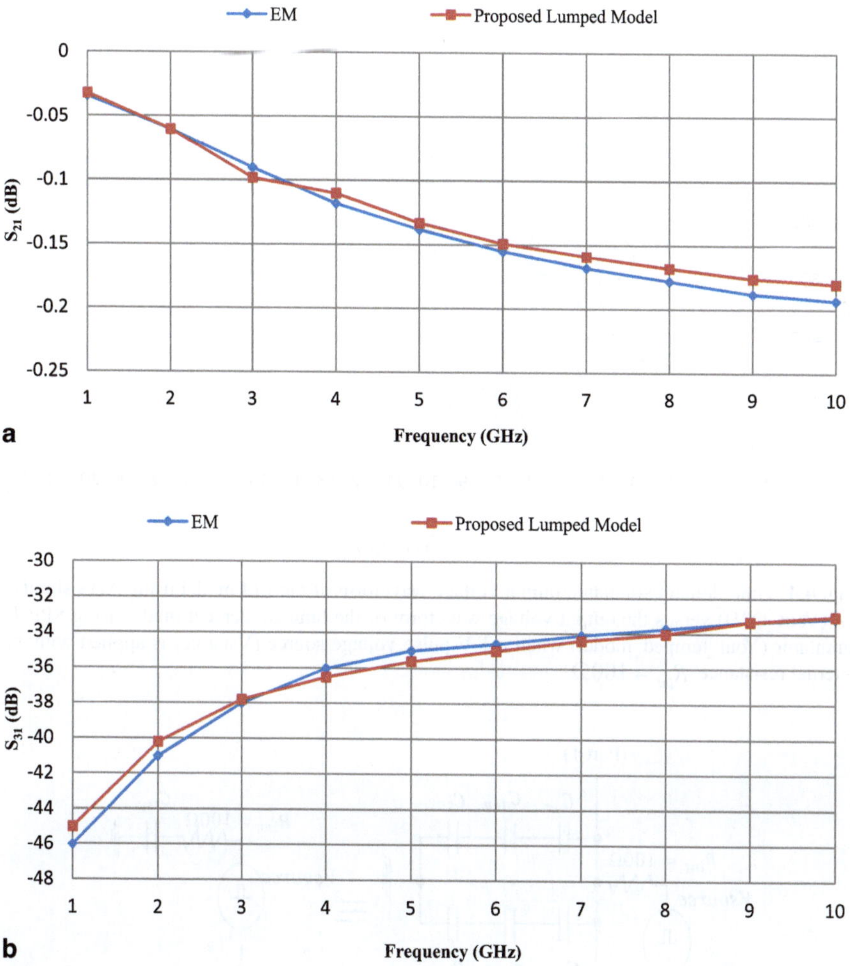

Fig. 4.3 A comparison of the quasi-static simulations against the lumped element model simulations for two linearly arranged TSVs (a) S_{21} parameters, (b) S_{31} parameter ($r_{tsv} = 5\mu m$, $l_{tsv} = 50\mu m$, $V_{tsv} = 0.3Vm$, $\omega_{bc} = 5\mu m$, $p_{tsv} = 15.6\mu m$)

to SPICE simulation [3] of the lumped model, where it is shown that both models produce almost identical waveforms (Fig. 4.4).

4.2 Studying the TSV Impact on Circuit Performance

The lumped element model of the TSV is simulated using SPICE to evaluate its impact on the 3D IC performance. A pulse voltage source, with a 100Ω internal resistance, is applied to the actual model of the cylindrical and the resultant waveform was

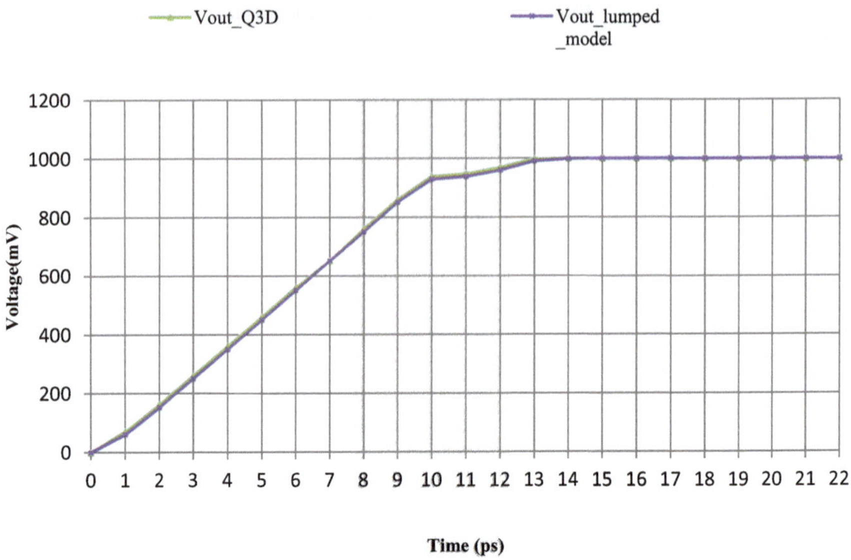

Fig. 4.4 Time-domain simulated output voltage waveform of the EM model using ADS simulator (Vout_Q3D) versus the output voltage waveform of the lumped element model using SPICE simulator (Vout_lumped_model) when a 1-V pulse voltage source (Vsource) is applied with an internal resistance $R_{int} = 100\Omega$

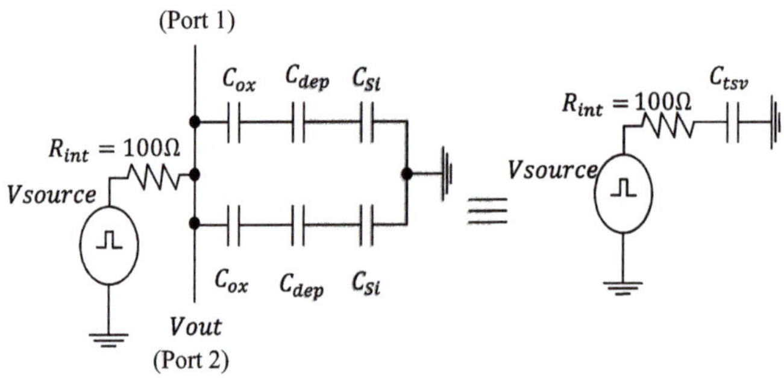

Fig. 4.5 Simplified circuit model for the TSV with all inductors short-circuited and all resistors open-circuited

compared against the simplified model shown in Fig. 4.5 (where all inductors and resistors of the metal are short-circuited and all resistors of silicon are open-circuited, i.e., the TSV metal is approximated as a perfect conductor neglecting the ohmic loss and skin effect). This simulation was performed using the ELDO simulator [3].

From the waveforms shown in Fig. 4.6, the error between the complete model and the simplified model is small; hence, the simplified model can be used in this case. Another experiment was performed for the case of $R_{int} = 10\Omega$ (i.e., representing a low-impedance device) where the considerable performance difference, shown in Fig. 4.7, recommends the use of the actual model. Hence, for high-impedance devices

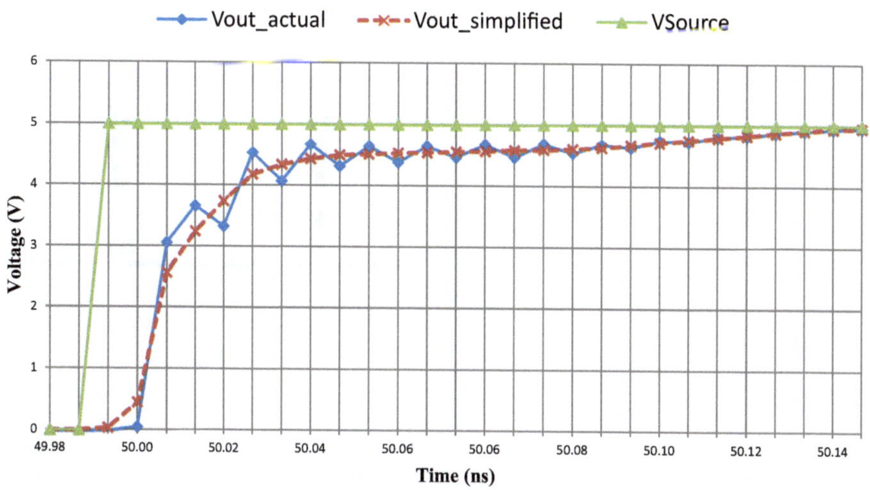

Fig. 4.6 Simulated voltage waveforms of the actual output voltage (Vout which is measure from port 2 in Fig. 4.1) versus the output from the simplified model (Vout_simplified which is measured from port 2 in Fig. 4.5) when a 5v pulse voltage source (VSource) is applied with an internal resistance $R_{int} = 100\Omega$

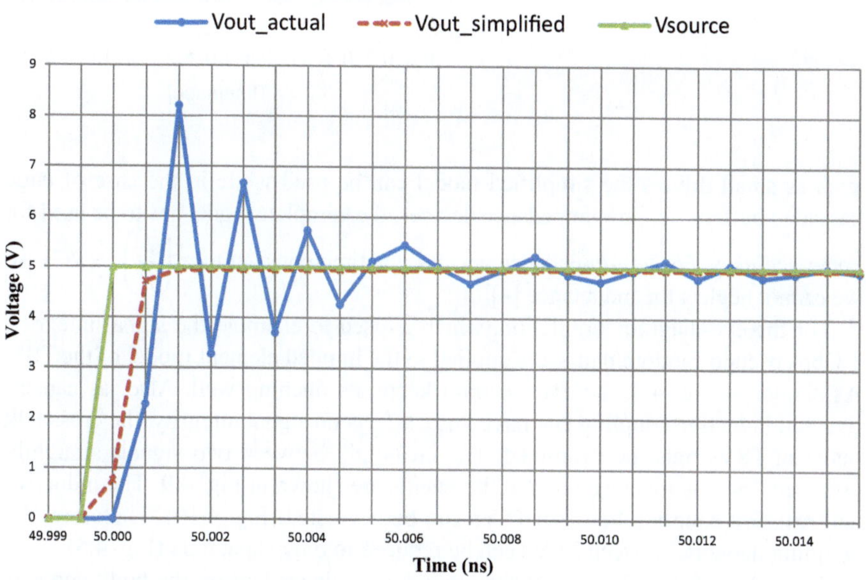

Fig. 4.7 Simulated voltage waveforms of the actual output voltage (Vout which is measure from port 2 in Fig. 4.1) versus the output from the simplified model (Vout_simplified which is measured from port 2 in Fig. 4.5) when a 5v pulse voltage source (Vsource) is applied with an internal resistance $r_{int} = 10\Omega$

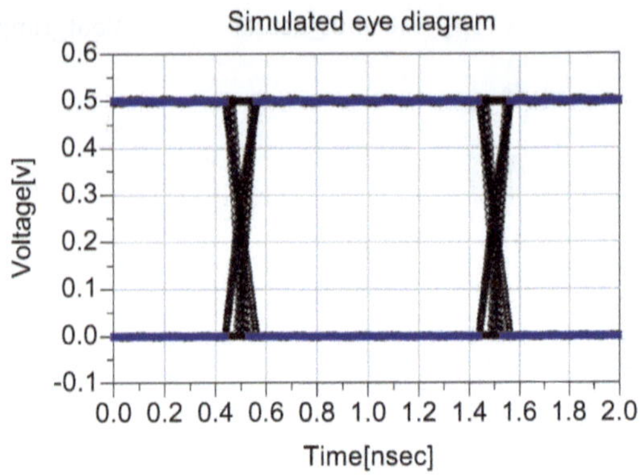

Fig. 4.8 Simulated eye diagram for the lumped element model (for a single TSV) at 1 Gbps

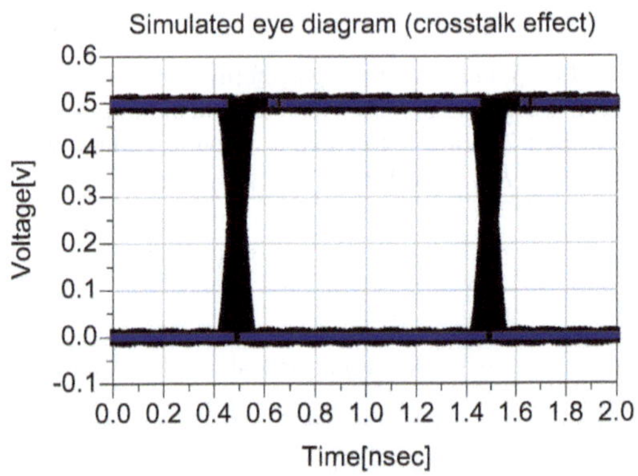

Fig. 4.9 Simulated eye diagram for the lumped element model (including cross talk effect between two TSVs) at 1 Gbps

such as small drivers the simplified model can be used while in the case of large cascaded buffers, i.e., low-impedance devices, the complete model has to be used for better accuracy. For low-impedance devices, the time-constant term $\left(\frac{L}{R}\right) > (RC)$, so we cannot neglect the inductance [4].

For the eye-diagram simulation, which is used to examine the signal integrity, 1 Gbps pseudo-random data were applied to the lumped element model of the TSV. As shown in Fig. 4.8, the eye diagram keeps its opening well. Also, as capacitive and inductive coupling can have large effects on signal integrity, the cross talk between TSVs must be examined. The cross talk between two signals straightly arranged TSVs is investigated and the results are shown in Fig. 4.9. The inductive and resistive coupling between TSVs can be neglected (Fig. 4.10). Therefore, the coupling network between TSVs can be reduced to only capacitors (Fig. 4.5).

The value of the silicon capacitance is highly dependent on the body contacts around the TSV. In the extreme case when there are no body contacts, this capacitance becomes similar to the stray capacitance, which is small (in the order of several fF).

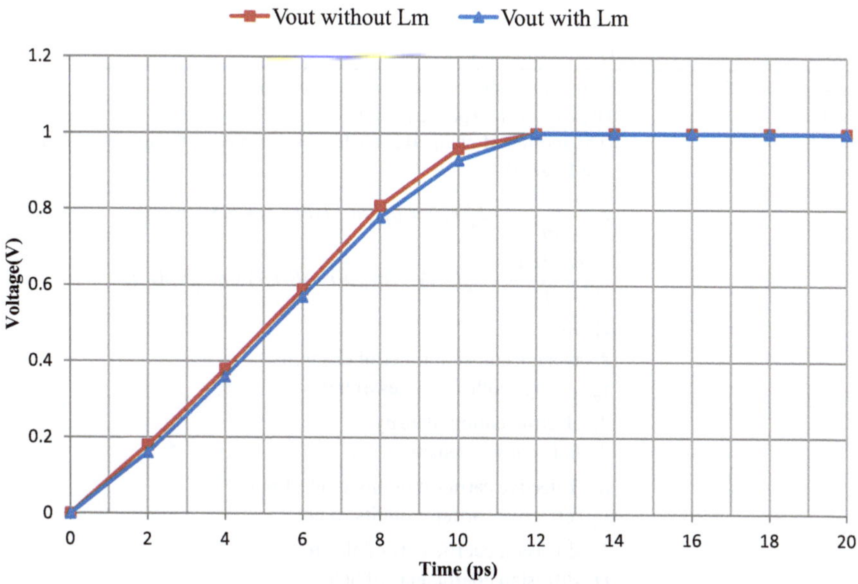

Fig. 4.10 Simulated voltage waveforms of the model between two TSVs (taking mutual inductance into consideration, but neglecting self-resistance and inductance) versus the output from the model (neglecting the mutual inductance effect, self-resistance, and inductance)

However, the presence of the body contacts around the TSV creates a virtual ground surface around the TSV. In this case, the silicon capacitance increases to several tens of fF. The more and closer the body contacts are to a TSV, the more the TSV capacitance and the isolation of the TSV from other circuit components.

4.3 Validation Against Device Simulation

The device simulator is a semiconductors-physics-based one, but it does not take into account magnetic field effects. Moreover, it is neglecting quantum effect (Table 4.1) [5–7]. The structure of the TSV was drawn then simulated on a device simulator. The used device simulator is the Sentaurus Device provided by Synopsys [8]. The Sentaurus Device is a multidimensional (1D/2D/3D) device simulator for silicon-based and compound semiconductor devices. The Sentaurus Device divides the simulated structure using an embedded meshing strategy taking into consideration the boundary conditions such as the electric field and the carrier currents.

After meshing, Sentaurus applies semiconductor equations and boundary condition on the meshed device structure by solving Poisson's equation and electron and hole continuity equations.

To simulate the TSV, a ground has to be defined. In VLSI, the ground is formed in a silicon wafer using body contacts deployed in the substrate at regular intervals

Table 4.1 Device simulator physics

Item	Device simulator
Physics-based theory	Semiconductors-physics-based simulators
Equations	The carrier transport equations (The motion of charged particles results in electric current: drift and diffusion) $\tau_n \frac{\partial J_n(r,t)}{\partial t} + J_n(r,t) = q\mu_n n(r,t)E(r,t) + qD_p \nabla n(r,t)$ $\tau_p \frac{\partial J_p(r,t)}{\partial t} + J_p(r,t) = q\mu_p p(r,t)E(r,t) + qD_p \nabla p(r,t)$, where τ_n average collision times of electrons τ_p average collision times of holes J_n electron current density J_p hole current density μ_n effective carrier mobility of electrons μ_p effective carrier mobility of holes D_n diffusion coefficients of electrons D_p diffusion coefficients of holes n electron density p hole density E electric field q electron charge
Disadvantages	1. Does not take into account particle theory of electromagnetic radiation (quantum effect) 2. Does not take into account magnetic field effects

of distance. Consequently, four arbitrary body contacts were used throughout the device simulations with different combinations of the body contacts being existent or not.

The dashed contour around the TSV structure represents the arbitrary positions at which we can place grounded body contacts. Simulations were done using different configurations of the body contacts through varying their number and the distance between them and the TSV. Simulations showed that the total TSV capacitance increases with the increase in the body contacts count. This increase is due to the fact that the more body contacts, the more shielding of the electric field lines emerging from the TSV.

Moreover, reduction in the distance separating the body contacts and the TSV increases the total TSV capacitance, because the closer the bodies contact, the more shielding it provides (Fig. 4.11). The closed-form expressions of the TSV lumped element model are examined against the device simulations and this comparison is shown in Fig. 4.12, where the comparison showed a good correlation and an error <2 % for the TSV capacitance.

Fig. 4.11 The simplified coupling network between TSVs, where $C_{ct} = (C_c \parallel 0.25C_c)$

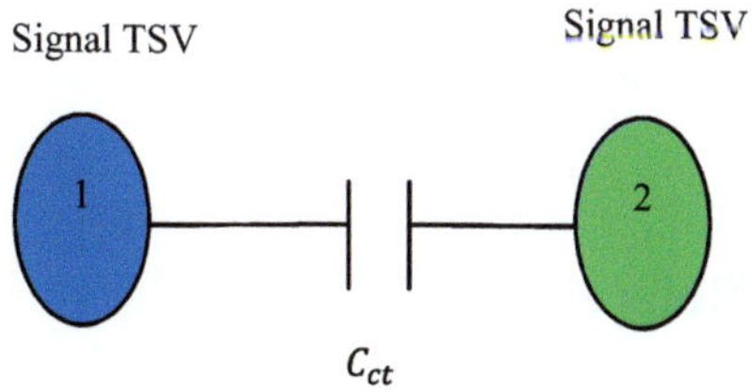

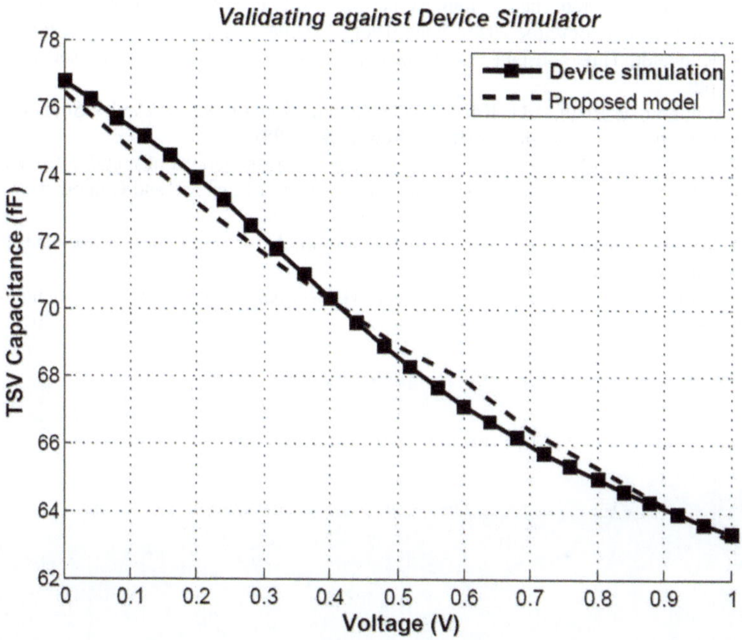

Fig. 4.12 A comparison of the device simulations against the lumped element model simulations

4.4 Summary

This chapter discussed the TSV lumped element model verification. The closed-form formulae achieved a very good agreement with the field solver scattering (S) parameter with error $<5\%$. The model was used to study the impact of a typical TSV on a 3D circuit performance. This study showed that the self-resistance and inductance of the TSV has negligible effect on high-impedance device characteristics; hence, in such a case the TSV can be considered a capacitance-dominated structure. Also, this study showed that the TSV self-inductance and resistance have a significant effect on low-impedance device characteristics. Also, it is shown that the inductive and resistive coupling between TSVs can be neglected and the coupling network between TSVs can be reduced to only capacitors. Furthermore, the TSV model was verified using device simulations with error $<2\%$.

References

1. Ansys, (n.d.), http://www.ansoft.com/products/si/q3dextractor. Accessed 2013
2. Agilent Technologies, (n.d.), http://www.home.agilent.com/agilent
3. Mentor Graphics, Eldo classic: foundry certified SPICE accurate circuit simulation (n.d.), http://www.mentor.com/products/ic_nanometer_design/analog-mixed-signal-verification/eldo
4. T. R. Kuphaldt, Time constant calculations (n.d.), www.ibiblio.org/kuphaldt/socratic/output/time2.pdf
5. L.H. Bomholt, P. Regli, T.O. Korner, W. Fichtner, *Combination of Device Simulation with Computational Electromagnetics* (IEEE, New York, 1998)
6. G. Wang, Coupled electromagnetic and device level investigations of metal–insulator–semiconductor interconnects, Ph.D. dissertation, Department of Science and Computers, Stanford University, 2001
7. F. M. Rotella, Equations for mixed circuit/device stimulation (n.d.), http://www-tcad.stanford.edu/~fmr/research/mixedmode/equations.html
8. Synopsis Inc., *Sentaurus Device User Guide, Ver. A-2008.09* (Synopsis Inc., Mountain View, 2008 Sep)

Chapter 5
TSV Macro-Modeling Framework

Macro-modeling of through silicon via (TSV) interconnects plays an important role in signal integrity (SI), power integrity (PI), and timing analysis of large 3D digital systems by reducing the CPU memory consumption and simulation time requirement. Hence, modeling methodologies are required to perform these simulations efficiently. This chapter presents a complete set of self-consistent equations including self and coupling terms for capacitance of general multi-TSV structures. The error when using the closed-form expressions as compared to a field solver is $<7\,\%$. As TSV parasitic capacitance is less than other conventional input/output (I/O) structures' capacitance; therefore, TSV technology results in lower I/O power consumption, which makes it suitable for low-power applications.

5.1 TSV Macro-Modeling Framework: Analysis and Verification

The main strategy that we follow here is the subdivision of the complex structure into separate and well-defined substructures. Each of these substructures is characterized separately by computing a macro-model, i.e., a set of equations that are able to reproduce the through silicon via (TSV) behavior. All the macro-models are then translated into an equivalent lumped element circuit, in order to connect them and to run a transient analysis using a standard Simulation Program with Integrated Circuit Emphasis (SPICE)-compatible circuit solver.

The subdivision is based on applying the sliding window technique to a numerical sequence that represents the TSV arrangement. It stops when the sliding window exceeds the lower and upper boundaries of the input vector.

In this section, we will propose a model that represents an arbitrary arrangement of a TSV array. In a TSV bundle, the self and coupling capacitance between each and every TSV can be defined by [1–4]:

K. Salah et al., *Arbitrary Modeling of TSVs for 3D Integrated Circuits*,
Analog Circuits and Signal Processing, DOI 10.1007/978-3-319-07611-9_5,

$$C_{bundle} = \begin{bmatrix} C_{1,1} & -C_{1,2} & -C_{1,3} & \cdots & -C_{1,n} \\ -C_{2,1} & C_{2,2} & -C_{2,3} & \cdots & -C_{2,n} \\ \cdot & & & & \cdot \\ \cdot & & & & \cdot \\ \cdot & & & & \cdot \\ -C_{n,1} & -C_{n,2} & -C_{n,3} & \cdots & C_{n,n} \end{bmatrix} \quad (5.1)$$

The electromagnetic (EM) simulations for various TSV arrangements show the following points:

1. If the number of surrounding TSVs is larger than four, the other surrounding TSVs effect can be neglected.
2. We should take into consideration whether each TSV is a signal or ground, where the stray capacitance increases when the surrounding TSV is a ground not a signal.

The equation for self capacitance for any TSV surrounded by an arbitrary number of other TSVs (on a grid) were found using curve fitting to electromagnetic simulations and to be given by:

$$C_S = C_{i,i} = C_{S0} + C_{C0}\,(0.75\,\alpha_1\beta_1 + 0.1875\,\alpha_2\beta_2 + 0.0938\,\alpha_3\beta_3 + 0.0938\,\alpha_4\beta_4) \quad (5.2)$$

where C_{S0} denotes the capacitance of the silicon substrate layer, C_{C0} corresponds to the coupling capacitance effect on the stray capacitance, and their expressions are deduced by the same way and are given by:

$$C_{S0} = \frac{0.007\,\pi\,(8n_{bc}+1)\,\varepsilon_0\,\varepsilon_{si} l_{tsv} \ln\left(2.7 + \frac{r_{tsv}+t_{ox}}{p_{tsv}}\right)}{\ln\left(1 + \frac{\omega_{bc}}{r_{tsv}}\right)}, \quad (5.3)$$

$$C_{c0} = 0.0045\,\pi\,\varepsilon_0\,\varepsilon_{si} l_{tsv} \ln\left(2.7 + \frac{r_{tsv}+t_{ox}}{p_{tsv}}\right) \quad (5.4)$$

$\alpha_1, \alpha_2, \alpha_3, \alpha_4$ are given in Table 5.1 and $\beta_2, \beta_3, \beta_4$ are given in Table 5.2. In the same way, an expression can be found for the coupling capacitance as follows:

$$C_c = C_{i,j} = \gamma_j\,\alpha_j\,\beta_j\,C_{c0} \quad (5.5)$$

where γ_j, α_j are given in Table 5.3 and β_j are given in Table 5.4.

To verify the basic idea of our algorithm to calculate the stray and coupling capacitance, an automated MATLAB code was written for this purpose, where we enter the TSV topology as an array, and each TSV is represented by two bits: The

Table 5.1 $\alpha_1, \alpha_2, \alpha_3, \alpha_4$ values

Number of neighboring TSV					
	0	1	2	3	≥4
α_1	0	1	1	1	1
α_2	0	0	1	1	1
α_3	0	0	0	1	1
α_4	0	0	0	0	1

Table 5.2 $\beta_1, \beta_2, \beta_3, \beta_4$ values

Neighboring TSV type		
	Signal TSV	Ground TSV
β_1	1	1.2
β_2	1	1.2
β_3	1	1.2
β_4	1	1.2

Table 5.3 γ_j, α_j values

Number of neighboring TSV					
	0	1	2	3	≥4
γ_j	0	1	1	1	1
α_j	0	1.2	1.1	0.9	0.8

Table 5.4 β_j values

Neighboring TSV type		
	Signal TSV	Ground TSV
β_j	1	0

first bit indicates whether the TSV is found or not while the other determine whether the TSV is signal or ground. The algorithm is applied to other different TSV arrangements and it was in a good agreement with the EM simulation results with only a 7 % difference (Table 5.5).

5.2 Multi-Stacked TSV

It is very difficult to obtain *S* parameters of multi-stacked TSV even though 3D full-wave EM solving method is used. Therefore, *S* parameters of multi-stacked TSV can be obtained by using *S* parameters of single TSVs and cascading them based on ABCD-parameter concept [5–7]. The main use of ABCD parameters is for chaining elements together, for example, if we want to obtain *S* parameter of two-stacked TSV, we cascade two *S* parameters of single TSV. From the z-parameter calculation by High Frequency Structure Simulator (HFSS), the ABCD parameters were obtained using the following equations:

Table 5.5 Percent error between simulation and closed-form expressions of TSV self- and coupling capacitance in different arrangements

Arrangement	Error percent for capacitance matrix (Err_c)

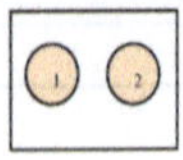

	1	2
1	1.2	1.7
2	1.7	1.6

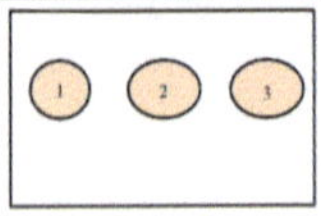

	1	2	3
1	1.1	4.3	3.8
2	4.3	5	3
3	3.8	3	2.5

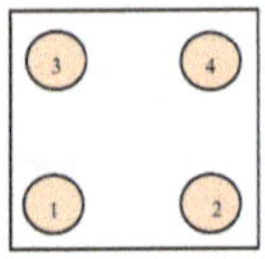

	1	2	3	4
1	1.3	6.2	4	5.2
2	6.2	5.1	1.5	2.9
3	4	1.5	3	3.1
4	5.2	2.9	3.1	6

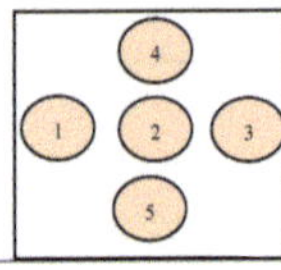

	1	2	3	4	5
1	1	3	1	4.3	4
2	3	6.2	4.4	3	2.1
3	1	4.4	2.7	1.05	3.7
4	4.3	3	1.05	5.1	1.9
5	4	2.1	3.7	1.9	4.35

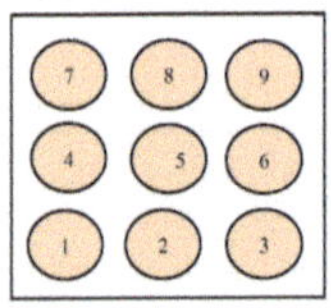

	1	2	3	4	5	6	7	8	9
1	1.35	2	2	4.2	2	4.2	2.1	3.8	2.2
2	2	5	5.2	3	4.1	1.1	6.3	4.2	1.5
3	2	5.2	1	4	3.4	2.1	4.2	3.7	5.7
4	4.2	5	4	2.9	1.1	1.16	2.3	5.1	1.6
5	1.9	4.1	3.4	1.1	2	4.1	2.2	4.3	1.35
6	4.2	1.1	2.1	1.16	3.2	5.5	3.23	5.1	4.3
7	2.1	6.3	4.2	5.2	2.2	6.3	3.4	2.3	5.2
8	3.8	4.2	2.2	5.1	6	5.1	3.9	6.2	2.5
9	2.2	1.06	5.7	2.2	1.35	3.47	5.2	2.5	4.3

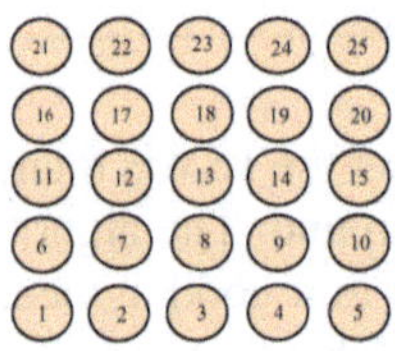

	1	2	3	4	5	6	7	8	9	10	11	12	13
1	1	3	6	4	5.1	1	3	1	4	5	6	4	4
2	3	3	6	3	1	1	3	3	5	1	6	5	3
3	6	6	1	4	5	6.5	4	3	1	4	5	3	6
4	4	3	4	5	1	4	3	1	1.1	1	6	5	2
5	5.1	1	5	1	4.3	2	4	4	2	4	3	5	6
6	1	1	3	4	2	2	1	6	6	6	3	4	1
7	3	4	4	3	4	4	3	3	7	5	6	2	3
8	6	3	5	1	1	6	1	3	1	1	4	4	1
9	4	5	1	4	2	3	7	1	3	5	5.1	6	4
10	5	1	1	1	6	6	5	1	5	2	2	7	3.6
11	6	1	5	4	3	3	6	4	5.1	3	4	4	2
12	6	5	6	5	5	4	2	4	3	7	4	5	2
13	4	3	6	2	6	1	3	4	4	3.6	2	2	2
14	4	5	4	3	2	3	2	2	2	7	3	4	5
15	2	6	5	6	1	5	3	4	3.8	6	1	4	6
16	2.4	3	4	6	2.2	6	3	1.7	4	5	3	3	5
17	6	5	2	6	1	3	6	5	2	1	3.1	6	2
18	1	4	3	3	1	3	6	5	2	4	2	1	1
19	4	4	1	3	3	6	2	2	2	6	5	4	5
20	3	3	2	2	6	2	1	1	5.3	2	2	3	5
21	2	3	2	4	6	5.6	3	2	5	5	5	3	3
22	5	4	4	6	2	2	7	2	3	5	5	4	4
23	4.8	3	5	4	3	1	4	4	4	5	7	4	4
24	1	3	6	4	3	1	1	4	1	4	2	2	4
25	2	3	6	1	2	2	4	2.9	4	3	2	1	2

	14	15	16	17	18	19	20	21	22	23	24	25
1	1	2	6.2	6	1	4	4	2	1	4.8	1	2
2	5	2	3	5	4	3	3	3	4	3	3	3
3	4	5	4	2	3	1	5	2	4	5	6	6
4	3	6	6	6	3	1	2	4	6	6	4	1
5	2	1	3	1	1	3	6	6	2	3	3	2
6	3	3	6	3	3	6	2	1.4	2	4	1	2
7	5	3	7	6	6	2	1	3	5	4	1	4
8	2	4	1.7	5	5	3	1	2	2	3	4	2.9
9	2	3.8	4	2	3	2	4	5	3	4	1	4
10	7	6	5	1	4	4.5	2	1	5	4	4	3
11	3	1	5.2	3.1	2	5	2	5	5	7	2	2
12	4	5	3	5	1	3	3	3	4	4	2	1
13	4	6	5	2	3	5	2	3	2	4	4	2
14	3	3	5	6	5	5	3	1.1	1	4	5	1
15	1	4	4	3	6	3	5	6.8	3	3	4	2
16	5	5	6	2	2	2	6	2.3	3	2	3	3
17	2	3	3	4	5	3	3	4	1.4	1	4	2
18	5	5	2	2	6	4	5	2	2	2	4.3	2
19	1	3	4	3	5	2	5	3	3	5.2	1	5
20	3	4.8	6	2	5	5	1	5	1	3	2	2
21	1	6.8	5	4	2	1	5	1	3	1	6	3
22	1	3	3	3	6	4	1	5	3	2.1	3	3
23	5	1	6	2	6	5.2	2	1	2.1	3	3	5
24	5	4	3	4	4.3	1	2	6	3	3	3	1
25	1	2	3	2	2	3	2	3	3	5	1	3

$$A = \frac{Z_{11}}{Z_{12}}, \tag{5.6}$$

$$B = \frac{Z_{11}Z_{22} - Z_{12}^2}{Z_{12}}, \tag{5.7}$$

$$C = \frac{1}{Z_{12}} \tag{5.8}$$

$$D = \frac{Z_{22}}{Z_{12}}. \tag{5.9}$$

The equivalent ABCD parameters of *n* multi-stacked TSV are given by the following equation:

$$[\mathrm{ABCD}]_{\mathrm{total}} = [\mathrm{ABCD}]_{\mathrm{tsv1}} * [\mathrm{ABCD}]_{\mathrm{tsv2}} * [\mathrm{ABCD}]_{\mathrm{tsvn}} \tag{5.10}$$

So, if we neglected metal coupling, the lumped model of *n* multi-stack TSVs is the cascading of *n* single TSV lumped model as shown in Fig. 5.1. Moreover, Fig. 5.2 shows the simulation results of the model versus the EM simulations for two-stacked TSVs which show good agreement.

Noting that there is no coupling between the TSV in layer 1 and the TSV in layer 2, as the metal layers ban the capacitive coupling and for the inductive coupling of both the TSVs as the current direction is vertical, the magnetic field is perpendicular to the page so they are parallel, and there is no magnetic coupling.

5.3 Summary

In this chapter, a complete set of self-consistent equations including self and coupling terms for capacitance of general multi-TSV structures was introduced. The error when using the closed-form expressions as compared to a field solver is <7 %. TSV parasitic capacitance is less than other conventional IO structures' capacitance; therefore, it is suitable for low-power applications.

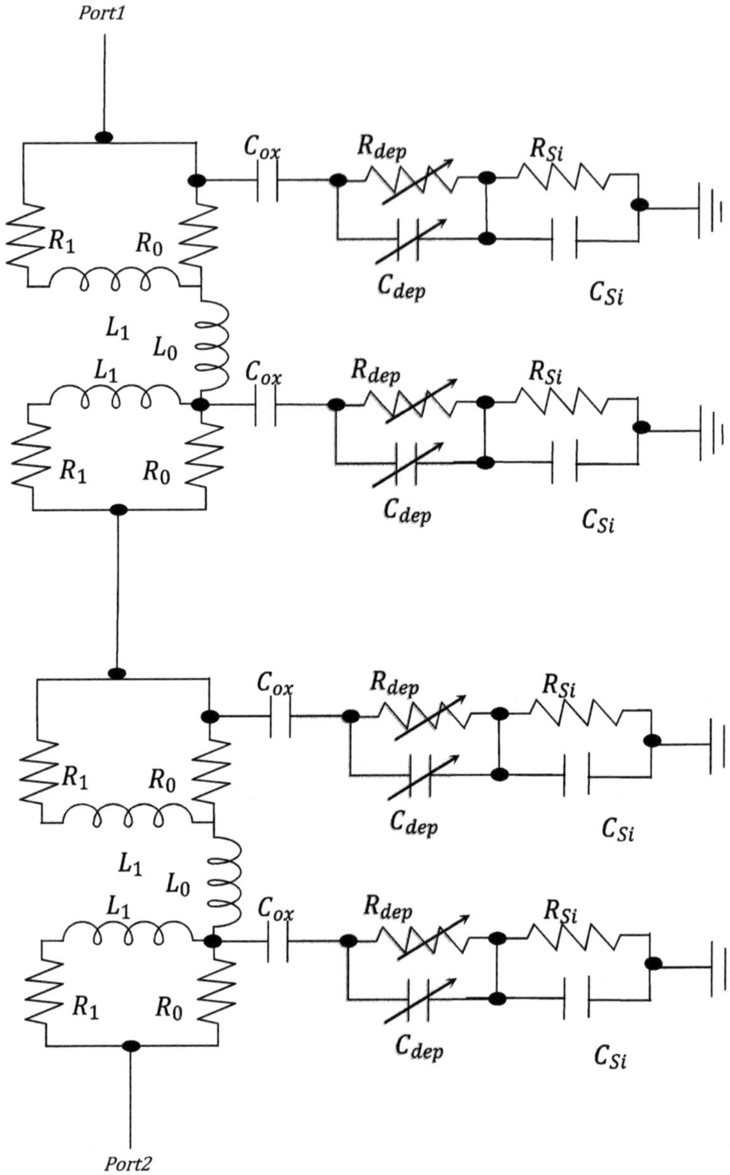

Fig. 5.1 Physical lumped circuit model for two-stacked TSVs

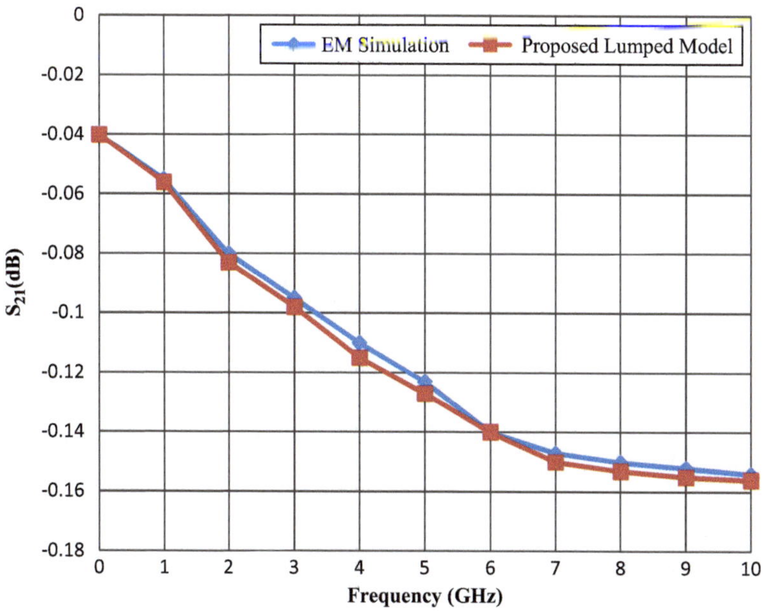

Fig. 5.2 A comparison of the EM simulations against the lumped element model simulations for two cascaded TSVs

References

1. R. Weerasekera, M. Grange, D. Pamunuwa, H. Tenhunen, On signalling over through-silicon via (TSV) interconnects in 3-D integrated circuits, DATE, 2010
2. M. Grange, R. Weerasekera, D. Pamunuwa, Optimal signaling techniques for through silicon vias in 3-D integrated circuit packages, in *Proceedings of IEEE Electrical Performance of Electronic Packaging and Systems*, 2010
3. H. Sane, Power supply noise analysis for 3D ICs using through silicon via, Master's Thesis. ECE Department Georgia Institute of Technology, 2010
4. M. Grange, R. Weerasekera, D. Pamunuwa, H. Tenhunen, Exploration of through silicon via interconnect parasitics for 3-dimensional integrated circuits, DATE, 2009
5. D. Chen, W. Chiou, M. Chen, T. Wang, K. Ching et al., Enabling 3D-IC foundry technologies for 28 nm node and beyond: through silicon-via integration with high throughput die-to-wafer stacking, in *Proceedings of IEEE International Electron Devices Meeting*, Dec 2009, pp. 1–4
6. J.S. Pak, C. Ryu, J. Kim, Multi-stacked through-silicon-via effects on signal integrity and power integrity for application of 3-dimensional stacked-chip-package, URSIGA, 2008
7. C. Ryu, J. Lee, H. Lee, K. Lee, T. Oh, J. Kim, High frequency electrical model of through wafer via for 3-D stacked chip packaging, *Electronics System-integration Technology Conference*, Sept 2006, vol. 1, pp. 215–220

Chapter 6
TSV Design Applications: TSV-Based On-Chip Spiral Inductor, TSV-Based On-Chip Wireless Communications, and TSV-Based Bandpass Filter

This chapter presents a novel design and characterization of the spiral inductor based on through silicon via (TSV) technology. An equivalent lumped model is introduced where this model is based on physics and takes into consideration TSV nonlinearities and skin effect.

A highly accurate closed-form expression for the TSV-based spiral inductor equivalent inductance is presented. This closed form is the first in literature. Moreover, this form is verified against a large number of electromagnetic (EM) simulations for different setups and shows excellent agreement with < 5 % error.

Also, an analytical equation to calculate the maximum quality factor is derived. According to the simulation results, the TSV-based spiral inductor exhibits better quality factor than a planar on-chip spiral inductor (120 % improvement) and a three-dimensional (3D) via-based spiral inductor (76 % improvement) for identical inductance.

Moreover, the self-resonance frequency of the TSV-based 3D spiral inductor is 38 % higher than the conventional two-dimensional (2D) spiral inductor and 3 % higher than the 3D via-based inductor. The inductor occupies only 15 % of the area of the conventional 2D planar spiral inductor and 60 % of the area of 3D via-based spiral inductor for the same inductance and higher quality factor. This small area of the inductor leads to significantly reducing the cost of the radio frequency system on chip (RF-SoC).

Moreover, this chapter presents a novel wireless TSV communication structure based on near-field inductive coupling. The system uses a spiral inductor built using TSV technology. The electrical performance obtained from EM-based *S*-parameter simulations for different number of turns and configurations is presented. The wireless TSV shows good coupling coefficient, and a closed-form expression for the coupling coefficient is presented. This coupling expression was verified against EM simulations and showed good agreement. The wireless TSV system provides a small area (30 μm)^2 for typical design values. This communication system is enabling network on chip (NoC). Moreover, under different supply voltages, there is no need for level shifters.

Moreover, in this chapter, we introduce a resonant inductive coupling for 3D integration circuits (3D-ICs) to enhance power transfer efficiency, where the res-

K. Salah et al., *Arbitrary Modeling of TSVs for 3D Integrated Circuits*, Analog Circuits and Signal Processing, DOI 10.1007/978-3-319-07611-9_6,

onance of an inductively coupled system increases the amount of magnetic flux linked between coils and improves the power transmission significantly.

Also, this chapter presents a new architecture for on-chip bandpass filters, based on TSV technology. This architecture is the first in literature. According to the simulation results, the TSV-based bandpass filter has an insertion loss of 1.5 dB at 90 and 20 GHz passband from 80 to 100 GHz. This band can be tuned by changing the dimensions of the coupled TSVs. The bandpass filter size is $180 \times 120 \times 300$ μm. This small area of the bandpass filter leads to significant reduction in the cost. This filter is suitable for mmW/THz applications.

6.1 Introduction: TSV-Based On-Chip Spiral Inductor

Inductors are considered important elements in analog, radio frequency (RF), and microwave circuits such as low-noise amplifiers, power amplifiers, filters, oscillators, impedance matching networks, and DC–DC converters [1]. Poor characteristics of integrated on-chip inductors are due to lossy characteristics of the substrate and high-resistance thin metal layers. Moreover, its large die area causes some limitations on routing. Therefore, integrated inductors are limiting factors in realizing radio frequency system on chip (RF-SOC) circuits [2]. The benchmarks for a spiral inductor are the quality factor and the self-resonance frequency (SRF) [3–6]. These benchmarks are summarized in Table 6.1. In most applications, the quality factor determines the performance of the circuit. For example, the phase noise of an LC oscillator is inversely proportional to the oscillator quality factor [7]. Therefore, it is important to obtain a high-quality factor inductance.

Besides the quality factor, the SRF is an important factor in inductors. At operating frequencies higher than the SRF, the impedance of the inductor becomes capacitive. Therefore, the maximum operating frequency of the circuit is limited by the SRF [8]. Hence, increasing the SRF is necessary to increase the bandwidth.

Two-dimensional (2D) planar and three-dimensional (3D) multilevel (via-based) spiral inductors have been reported [9–12], but they have very small inductance values while consuming large areas. In general, there is a trade-off between the Q-factor and area. Therefore, a spiral inductor with high Q-factor is still a challenge [13].

Three dimensional integrated circuit (3D-IC) is now considered as a new paradigm in IC design promising to overcome the limitations of 2D-ICs, reduce the interconnect bottleneck, improve performance, and enable heterogeneous stacking of different layers like RF, analog, digital, and microelectromechanical systems (MEMS) into a single 3D-IC (Fig. 6.1). Today, many 3D stacking technologies exist, but through silicon via (TSV) is one of the key points and is considered an excellent candidate for 3D integration for different device layer connection.

TSV-based 3D integration is a promising solution to improve the quality factor and spiral SRF over planar 2D spiral (Fig. 6.2) and 3D via-based spiral (Fig. 6.3),

Table 6.1 Benchmarks for a spiral inductor

Benchmark	Description
The quality factor (*Q*)	The quality factor is defined as a ratio of the net energy stored by magnetic field and the average power dissipation in one cycle, and calculated as: $$Q = 2\pi.\frac{\text{Energy stored}}{\text{Energy loss in one oscillation cycle}} = 2\pi.\frac{\text{peak magnetic energy} - \text{peak electric energy}}{\text{Energy loss in one oscillation cycle}},$$ Q is decreasing at a certain frequency range, where the capacitive effects dominate the behavior of the spiral. Also, the substrate loss at high frequencies can degrade the quality factor, where the parasitic resistance consumes the stored energy and the parasitic capacitance reduces the inductivity
The self-resonance frequency (*SRF*)	Due to the extra parasitic capacitances, a spiral inductor has a SRF when the average electric and magnetic energies are equal. Therefore, its factor will vanish to zero at (SRF). The SRF can be calculated as: $$SRF = \frac{1}{2\pi\sqrt{L_{eq}C_{eq}}},$$ where L_{eq} and C_{eq} are the equivalent inductance and capacitance of the inductor respectively. Obviously, the larger the parasitic capacitance, the lower the SRF

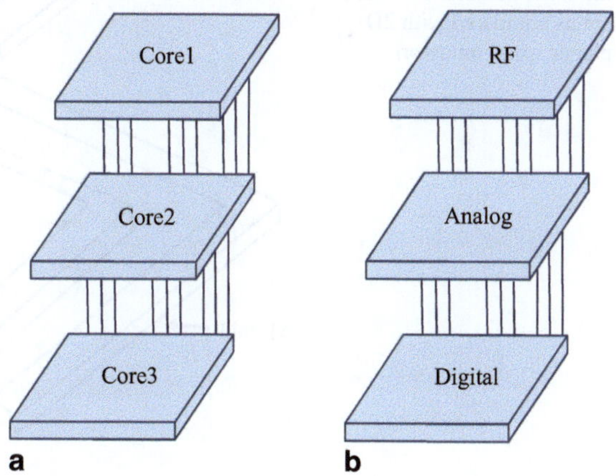

Fig. 6.1 In the 3D design, a 2D chip is divided into a number of different blocks and each one is placed on a separate layer of silicon where each layer is stacked on top of each other: **a** multi-core processors stacking and **b** heterogeneous stacking (digital, analog, RF)

where it reduces parasitics and losses, which are the main causes for the quality factor degradation.

Comparing to a 2D spiral, for the same footprint, the resistance in case of 3D TSV-based spiral is much lower. Therefore, the quality factor is better. This is due to the fact that a TSV has a controllable and large cross section area as compared

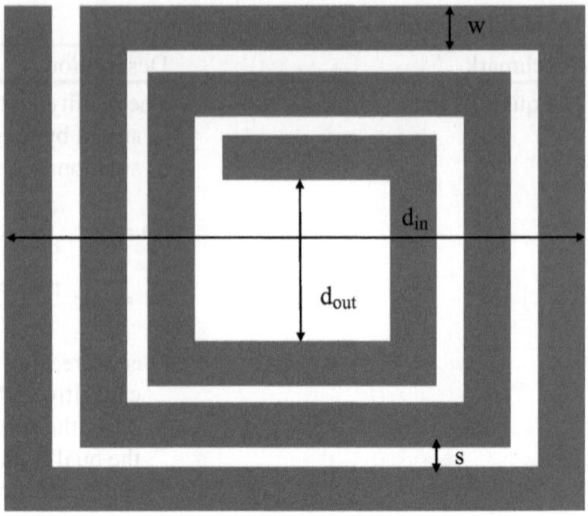

Fig. 6.2 The typical layout for the 2D spiral inductor. The 2D planar spiral inductor can be defined by the design parameters, which are the outer/inner diameter d_{out}/d_{in}, the metal width *w*, the spacing between the wiring metal *s*, and the number of turns *n*

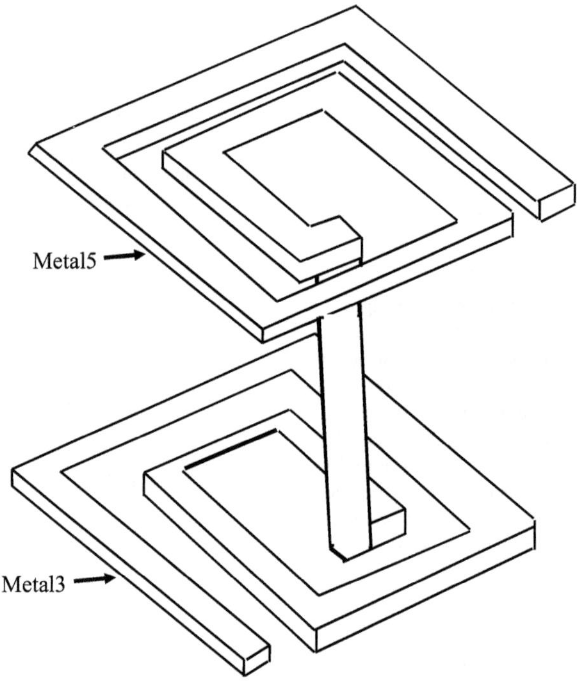

Fig. 6.3 The typical layout for the 3D via-based spiral. It consists of series-connected spiral inductors in different metal layers (stacked inductors). The wires wind from the top metal layer to downward the bottom one. The metal thickness increase upward. They can generate the same inductance in less are as compared with 2D planar spiral inductors

to a 2D spiral, which is constrained by the metal layer thickness that is fixed by a standard complementary metal-oxide semiconductor (CMOS) process technology. (The illustration can be demonstrated in Table 6.2).

As compared to a 3D via-based spiral, which uses the multiple metal layers to achieve the required inductances in a small area, a 3D TSV-based spiral uses less area and less parasitic resistance since the 3D via structure still uses the thinner lower metal layers.

Table 6.2 Current direction in case of 2D planar spiral inductor and 3D TSV-based spiral inductor

	2D conductor	3D conductor
Current direction 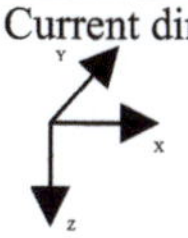		
The resistance	$R_{2D}\ \alpha\ \frac{y_2}{x_2 z_2}$ (6.1)	$R_{3D}\ \alpha\ \frac{z_3}{x_3 y_3}$ (6.2) (for simplicity, we treat the circular TSV as a square one)
From Eqs. (6.1), (6.2) and assuming that $y_2=z_3$ and $x_2=x_3$: $\frac{R_{3D}}{R_{2D}}=\frac{z_2^{\,2}}{y_3^{\,2}}$, (6.3) where z_2 (the metal thickness) is fixed by the CMOS process and y_3 is controllable and could be less than z_2 so $R_{3D} < R_{2D}$		

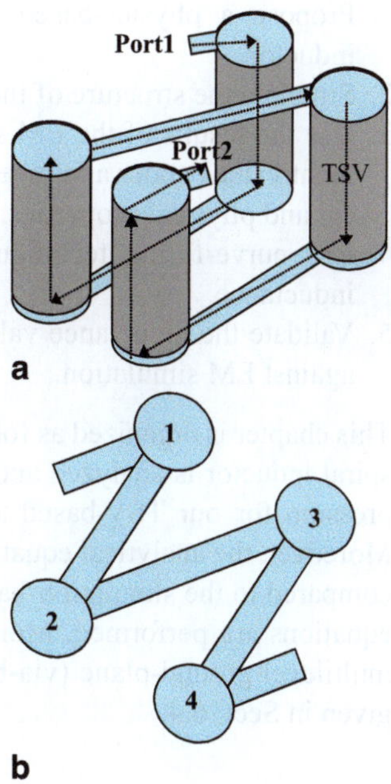

Fig. 6.4 Schematic of two turns TSV-based spiral inductor: **a** 3D view and **b** top view. The fundamental difference between TSVs and vias is the semiconducting material surrounding TSVs

This work introduces the TSV interconnect paradigm as a new architecture for the spiral inductor. Figure 6.4 shows the concept of the winding of the TSV-based inductor, where each turn is modeled as vertical TSV.

The spiral inductor parameters and their typical range are defined in Table 6.3, where we use a wide range of inductors, with TSV diameter (D_{tsv}) varying from 5

Table 6.3 TSV-based spiral inductor typical values, where a wide range of inductors are used

Design parameters	Typical range	Typical design value
Number of turns (N)	1–3	3
TSV diameter (D_{tsv})	5–100 μm	30 μm
TSV length (l_{tsv})	50–300 μm	300 μm
Metal track width (w_m)	5–100 μm	5 μm
Metal track length (l_m)	20–100 μm	20 μm
Metal track thickness (t_m)	1–5 μm	1 μm
Space between two metal tracks (P_m)	10–100 μm	10 μm

to 100 μm, TSV length (l_{tsv}) varying from 50 to 300 μm, metal track width (w_m) varying from 5 to 100 μm, metal track length (l_m) varying from 20 to 100 μm, metal track thickness (t_m) varying from 1 to 5 μm, space between two metal tracks (P_m) varying from 10 to 100 μm, and number of turns (N) varying from 1 to 3. Also, from Fig. 6.4, it is noticed that the N-turns 3D spiral inductor consists of $2N$ TSVs.

The methodology used in the modeling of the TSV-based spiral can be summarized in the following steps:

1. Propose a physics-based lumped element model for the TSV-based spiral inductor.
2. Simulate the structure of the TSV-based spiral inductor using EM simulation.
3. Use the results of the EM simulator to optimize the physics-based lumped-element values under a number of different setups (TSVs dimensions, TSV electrical and physical properties, and substrate parameters).
4. Use curve-fitting techniques, to develop a closed-form expression for the inductance.
5. Validate the inductance value set using the closed-form expressions of step (d) against EM simulation.

This chapter is organized as follows. In Sect. 6.2, the architecture for the TSV-based spiral inductor is analyzed and simulation results are discussed. A closed-form expression for our TSV-based spiral inductor equivalent inductance is also given. Moreover, the analytical equations for the maximum quality factor are derived to be compared to the simulation-based one. In Sect. 6.3, verification of the closed-form equations are performed. Moreover, a comparison with conventional on-chip and multilevel ground-plane (via-based) spiral inductors is presented. Conclusions are given in Sect. 6.4.

6.2 TSV-Based Spiral Architecture and Analysis

In this section, we will propose an equivalent lumped element model, discuss the EM simulation results, and deduce closed-form expressions for the inductance and the maximum quality factor.

Table 6.4 Geometric variants for the TSV-based spiral inductor and the associated estimated self-inductance, resistance, and quality factor values

N	*L*(nH)	*R*(Ω)	Q_{max}/*F*(GHz)	*SRF*(GHz)
1	1.2	1.26	21.2/5.2	18.1
2	2.3	2	12.1/2.7	16.5
3	3.5	2.9	8.12/1.8	15.1

6.2.1 The TSV-Based Spiral Inductor: Analysis and Characterization

As stated earlier, in this chapter we introduced a new architecture for the spiral inductor based on TSVs. Various TSV-based spiral inductors are designed and simulated using EM simulator Ansoft HFSS™. The simulated inductance, resistance, quality factor, and SRF at different number of turns are shown in Fig. 6.5 and Table 6.4, where we obtained peak quality factor ranging from 8 to 21 for inductance ranging from 1.2 to 3.5 nH.

The simulation results show the following points:

1. While increasing the number of turns of the TSV-based 3D inductor, more and more metal-to-metal and metal-to-substrate capacitances are generated. Thus, the equivalent capacitances become larger. Hence, the SRF decreases while the number of turns increases, i.e., smaller inductor area results in higher SRF.
2. It is noticed that a TSV-based spiral inductor is proportional to the area of the loop ($l_m \times l_{tsv}$) and the number of turns (N).
3. A smaller spacing improves the interwinding magnetic coupling and reduces the area consumed by the spiral. However, a large spacing is only desired to reduce the interwinding capacitance.
4. For a spiral inductor, a small inductance has a high Q, and vice versa, a large inductance has a low Q.
5. Increasing the metal line width reduces the series resistance, but increases the parasitic capacitance to substrate, which degrades both Q and SRF.

6.2.2 Lumped Element Model for the TSV-Based Spiral Inductor and Derivation of Q_{max}

To allow Simulation Program with Integrated Circuit Emphasis (SPICE) simulations (time-domain simulation), a lumped element model for the TSV-based spiral inductor is introduced (Fig. 6.6) where the interpretation of each element is demonstrated in Table 6.5. Those elements account for [14]:

1. Substrate losses: where the capacitive coupling allows conduction current to flow not only in the TSV but also through the substrate.

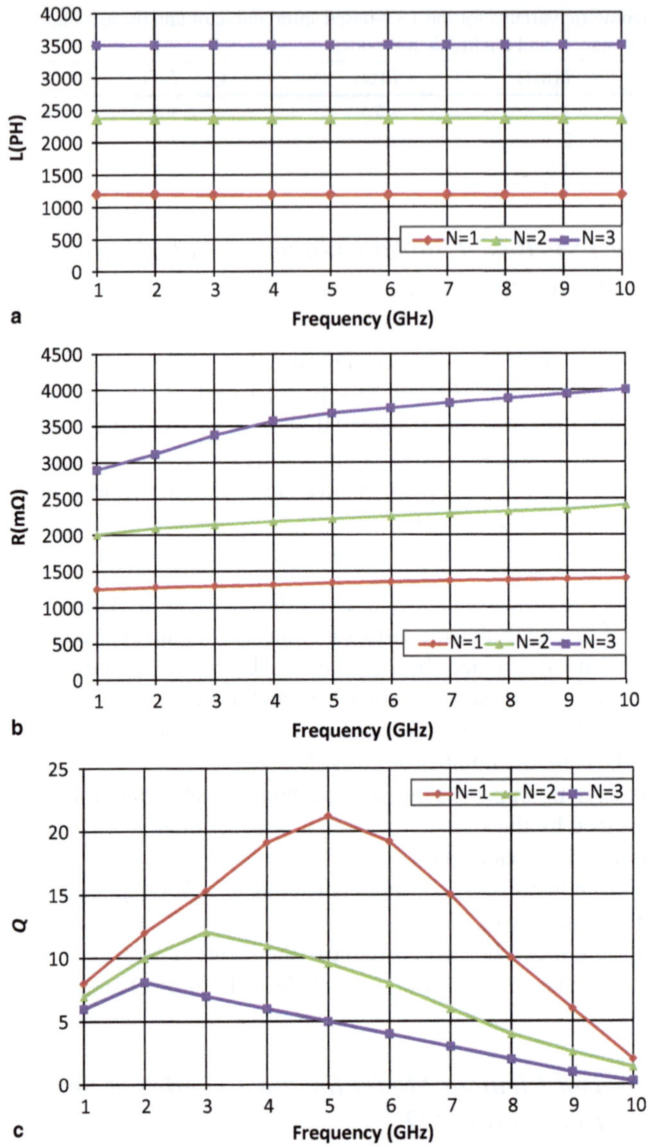

Fig. 6.5 EM simulation variations of **a** L, **b** R, **c** Q of 3D TSV-based spiral inductor with one to three turns versus frequency (1–10 GHz)

2. Metal losses: due to the copper resistance.

The model parameters of a particular spiral inductor could be estimated by a fitting procedure on the data provided by the EM simulations. The fitting procedure was

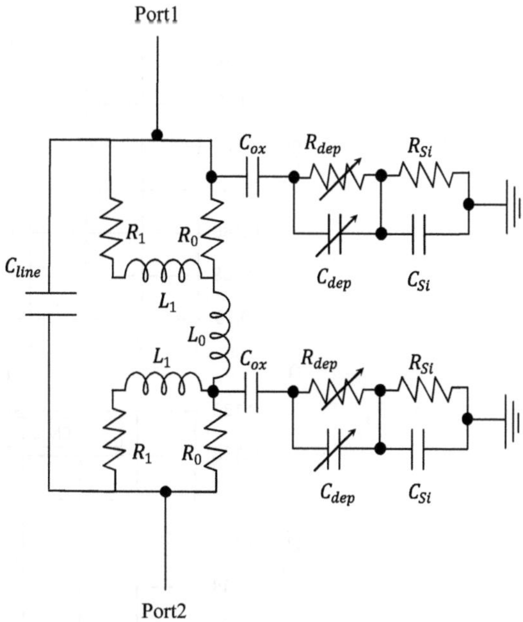

Fig. 6.6 The equivalent lumped model for TSV-based spiral inductor

Table 6.5 The model lumped elements and its physical interpretation

Circuit element	Physical meaning
R_0, L_0	Inductance and ohmic losses of the spiral
R_1, L_1	Skin effect of the spiral
C_{ox}	Capacitance of the oxide
C_{dep}, R_{dep}	Silicon substrate depletion region capacitance and resistance
C_{si}, R_{si}	Silicon substrate capacitance and resistance
C_{line}	Capacitance between two metal tracks of the spiral

performed by minimizing the least square error between the circuit element and EM data using curve-fitting software.

The capability of this model to describe the behavior of spiral inductors is investigated, where the *S* parameters obtained by the circuit model are compared with the ones provided by EM simulations for various test cases (Fig. 6.7). The comparison shows a good agreement and confirms that the physical model can indeed predict the overall inductor behavior.

The extracted values of the equivalent circuit models are reported in Table 6.6. From the obtained values and for simplicity, we will neglect R_{dep}, C_{dep}, C_{ox}, R_1, L_1, and the model can be simplified to the model shown in Fig. 6.8. This simplified model will help us to obtain closed-form expressions for the maximum quality factor.

In circuit applications, spiral inductors could be used in two possible configurations [15] (Fig. 6.9):

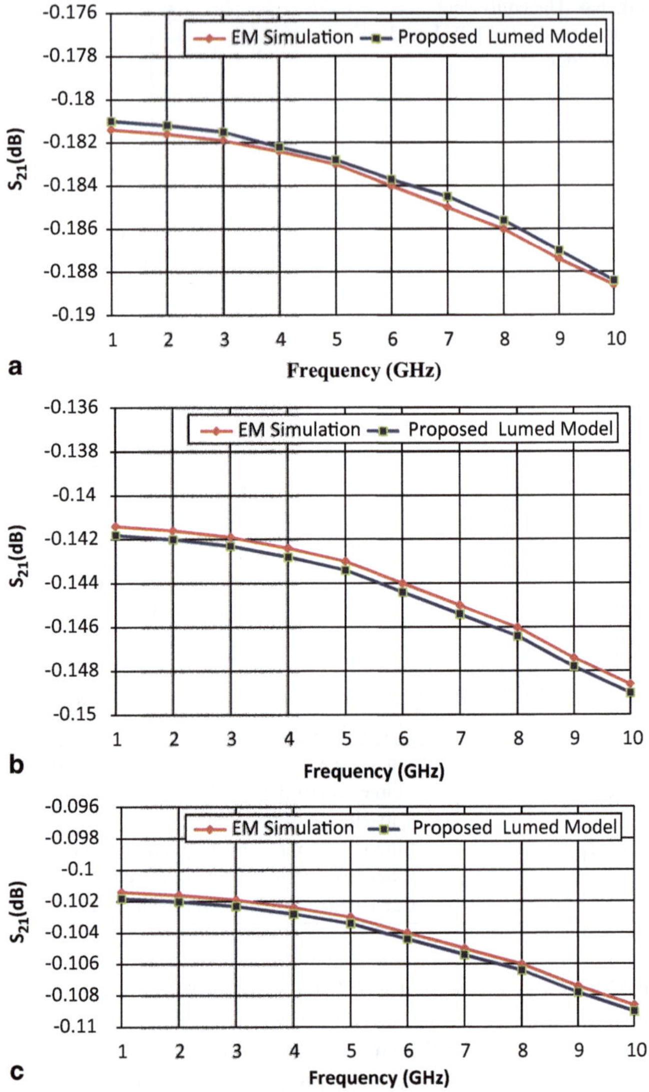

Fig. 6.7 A comparison of the EM simulations against the lumped element model simulations for **a** $N=1$, **b** $N=2$, **c** $N=3$

1. One port or single ended: when one of the two ports is shorted.
2. Two port: when neither of the two ports is shorted and is not symmetrically driven.

The quality factor for single-ended lumped model can be calculated as follow:

$$Q = \frac{im(Y_{11}^{-1})}{re(Y_{11}^{-1})}. \tag{6.4}$$

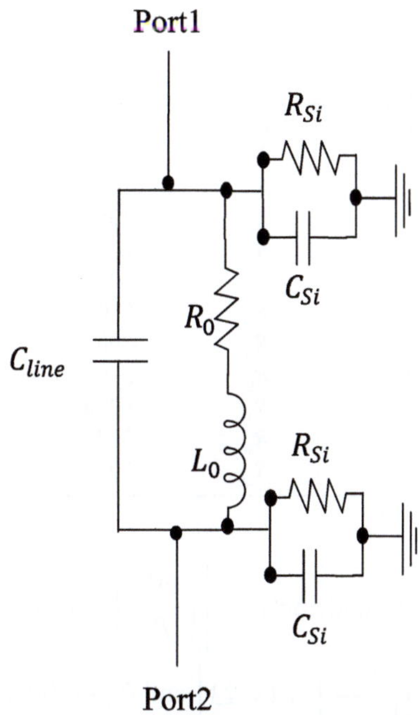

Fig. 6.8 Simplified model of the spiral inductor. This simplified model will help us to obtain closed-form expressions for the maximum quality factor

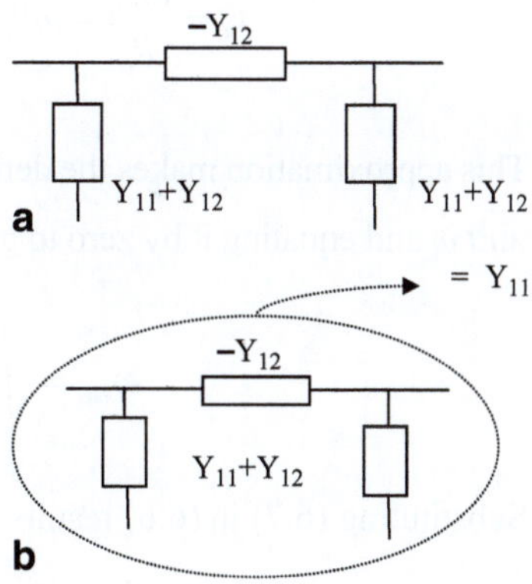

Fig. 6.9 Simplified equivalent model representation: **a** two-ports model and **b** one-port model

Using Eq. (6.4), the quality factor Q for the simplified model (Fig. 6.8) can be written as [16]:

$$Q = \frac{\omega L_0}{R_0} \frac{R_{si}\left(1 - \frac{R_0^2 (C_{line} + C_{si})}{L_0} - \omega^2 L_0 (C_{line} + C_{si})\right)}{R_{si} + \left[\left(\frac{\omega L_0}{R_0}\right)^2 + 1\right] R_0}$$

$$= \frac{\omega L_0}{R_0}. \text{ Substrate loss factor. Self resonance factor.} \quad (6.5)$$

Table 6.6 Extracted values for *R, L,* and *C* of the TSV-based spiral inductor model at (N=1, F=1 GHz)

N	$R(\Omega)$		C(fF)		L(nH)	
1	R_0	1.3	C_{ox}	122	L_0	0.98
	R_1	20 m	C_{si}	15	L_1	0.03
	R_{si}	1.1 K	C_{dep}	120		
	R_{dep}	105	C_{line}	200		
2	R_0	2.1	C_{ox}	202	L_0	2.1
	R_1	40 m	C_{si}	22	L_1	0.07
	R_{si}	2.3 k	C_{dep}	200		
	R_{dep}	300	C_{line}	403		
3	R_0	3.3	C_{ox}	300	L_0	3.3
	R_1	60 m	C_{si}	34	L_1	0.11
	R_{si}	3.5k	C_{dep}	320		
	R_{dep}	505	C_{line}	612		

By noticing that $\left[\left(\frac{\omega L_0}{R_0}\right)^2+1\right]R_0$ is much smaller than R_{si}, where at F=1 GHz, $\left[\left(\frac{\omega L_0}{R_0}\right)^2+1=23\right]$ and R_{si}=1100, the quality factor is approximated by:

$$Q=\frac{\omega L_0}{R_0}-\omega R_0(C_{line}+C_{si})-\frac{\omega^3 L_0^{\,2}(C_{line}+C_{si})}{R_0}. \tag{6.6}$$

This approximation makes the derivation $\frac{d}{d\omega}$ slightly easier. So, by derivation of Q *w.r.t* ω and equating it by zero to get the frequency at the maximum, Q results in:

$$\omega_{max}=\sqrt{\frac{1}{3}\left(\frac{1}{L_0(C_{line}+C_{si})}+\frac{R_0^{\,2}}{L_0^{\,2}}\right)}. \tag{6.7}$$

Substituting (6.7) in (6.6) results in:

$$Q_{max}=\frac{\omega_{max}L_0}{R_0}-\omega_{max}R_0(C_{line}+C_{si})-\frac{\omega_{max}^{\;3}L_0^{\,2}(C_{line}+C_{si})}{R_0}. \tag{6.8}$$

The Q_{max} values obtained from Eq. (6.8) and the values obtained from the simulations (Table 6.4) at N=1, 2, and 3 are correlated and their values are shown in Table 6.7.

6.2.3 *Closed-Form Expressions for the Inductance*

In this section, we will derive a closed-form equation for the inductance. The calculation of the inductance in spiral structures is formulated based on the self-

Table 6.7 Comparison between (q_{max}, ω_{max}) values obtained from Eqs. (6.8) and (6.7) and their values obtained from the simulations (Table 6.4) at $n=1$, 2, and 3

N	Q_{max}		
	Analytical value	Simulated value	Percentage error (%)
1	21	21.2	0.95
2	11.8	12.1	2.5
3	8	8.12	1.5
N	ω_{max}		
	Analytical value ($3\pi \times 10^9$ rad/s)	Simulated value ($3\pi \times 10^9$ rad/s)	Percentage error (%)
1	5.4	5.2	3.8
2	2.8	2.7	3.7
3	1.85	1.8	2.7

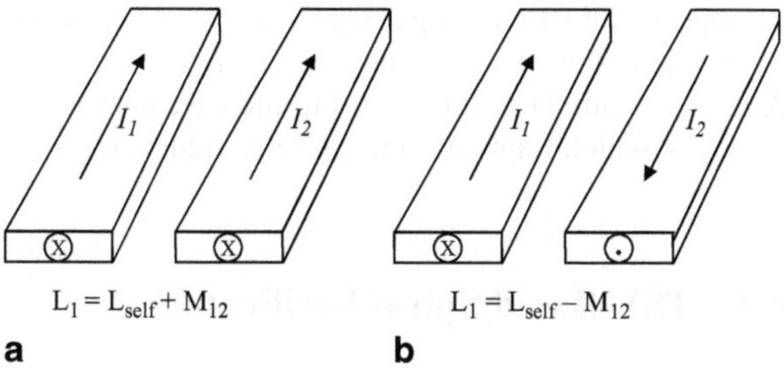

Fig. 6.10 **a** Parallel currents travelling in same directions contribute positive mutual components of inductance (even mode) and **b** parallel currents travelling in opposite directions contribute negative mutual components of inductance (odd mode)

inductance, which is fixed by the geometry of each line segment in the coil, and by the mutual components, which are determined by the geometry, the spatial separation, as well as by the relationship between the currents carried by those lines. Parallel currents travelling in the same direction contribute positive mutual components of inductance, whereas parallel currents travelling in opposite directions contribute negative mutual components of inductance (Fig. 6.10) [17–19].

Using 3D EM simulations along with curve-fitting techniques, the closed-form equation for our TSV-based spiral inductor equivalent inductance is given by:

$$L_{tsv\_spiral} = \frac{\mu_0 N^2 l_{tsv} l_m}{2\pi(ND_{tsv} + 0.1(N-1)p_m)}. \tag{6.9}$$

From the above equation, the inductance is a function of number of turns, metal track length, TSV diameter, TSV length, and spacing between tracks. This formula is scalable over a wide range of spiral dimension. From Fig. 6.11 [20, 21], we can notice the following to increase the inductance:

1. Decrease the distance between the conductor segments that carry currents in the same direction {(seg3, seg7), (seg5, seg1), (seg6, seg2)} to increase positive mutual inductance, i.e., decrease (P_m). In other words, the direction of current for the first loop is the same as the second one. Therefore, it is important to make them near each other.

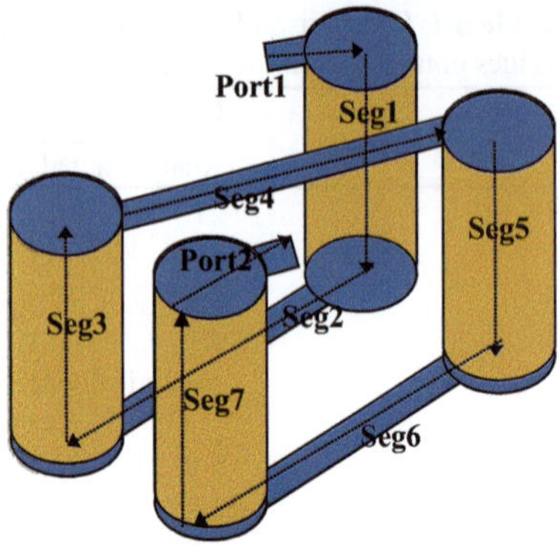

Fig. 6.11 The conductor segments numbering for the TSV-based spiral inductor

2. Increase the distance between the conductor segments that carry currents in the opposite direction {(seg3, seg1), (seg5, seg7), (seg6, seg4)} to decrease negative mutual inductance, i.e., increase (l_m, l_{tsv}).
3. Increase number of turns (N) to increase number of positive mutual inductance terms, which implicitly increase the inductance.

6.3 TSV-Based Spiral Verification

In this section, our lumped element model and closed-form formulae will be validated against EM simulations. Moreover, numerical comparisons with related work are presented.

6.3.1 Validation Against EM Simulation

The approach which is applied to validate the lumped model and closed-form equations presented above (see Sect. II.B, Sect. II.C) is to simulate new random physical parameter test cases using EM simulator and compare them with the lumped model and closed-form equation. For all those new random physical parameters, the S parameters of the EM simulations are correlated with the S parameter of the lumped element model whose inductance values are found using the closed-form expressions. The percent error between simulation and closed-form expressions of TSV-based spiral inductor, inductance ($L_{\text{tsv\_spiral}}$) are shown in Table 6.8, where the maximum error is less than 5 %.

6.3.2 Comparison with Related Work

From the comparison with related work, it is clear that the TSV-based spiral is offering large quality factor and high SRF, where the obtained Q from the TSV-based

Table 6.8 Percent error between simulation and closed-form expressions of TSV-based spiral inductor inductance (L_{tsv_spiral})

D_{tsv} (μm)	l_m (μm)	p_m (μm)	l_{tsv} (μm)			
			100	150	200	300
15	10	10	2.1	2.4	3.3	4.7
	100	50	1.9	4.2	2.6	3.8
20	10	10	2.7	4.2	2	2.7
	100	50	3.9	1.4	3.2	2.5
25	10	10	1.5	1.5	3	1.3
	100	50	3.2	4.8	2.7	1.8
30	10	10	3	3.1	4.2	4
	100	50	1.5	2	4.3	3.6
35	10	10	1.5	1.5	3.1	1.3
	100	50	3.2	4.8	2.7	1.8
40	10	10	3	3.1	4.2	4
	100	50	1.5	2	4.1	3.6
45	10	10	1.5	1.5	2	1.3
	100	50	3.2	4.8	2.7	1.8
50	10	10	2.7	3.3	4.2	4
	100	50	1.5	2	4.3	2.6
55	10	10	1.6	2.35	3	2.2
	100	50	1.8	3.5	2	2.3

Table 6.9 Comparison between planar spiral inductor, via-based spiral inductor, and TSV-based spiral inductor for the same inductance (0.88 nH, 2.5 nH) and frequency (1 GHz)

Parameters	Conventional planar spiral inductor		3D via-based spiral inductor	This work: 3D TSV-based spiral inductor
	[21]	[20]	[13]	
L(nH)	0.88	0.88	0.88	0.88
R(Ω)	0.92	1.1	0.73	0.527
Q	6	5	7.5	13.2
Area	–	–	(50 μm)^2	(35 μm)^2
SRF	–	–	–	10.5 GHz
Parameters	*Conventional planar spiral inductor*		*3D via-based spiral inductor*	*This work: 3D TSV-based spiral inductor*
	[8]		[8]	
L(nH)	2.5		2.5	2.5
R(Ω)	–		–	0.827
Q	6		6.8	11.8
Area	(320 μm)^2		(60 μm)^2	(30 μm)^2
SRF	11.8 GHz		15.8 GHz	16.3 GHz

"–" means it is not mentioned

inductor is higher than the values obtained from identical inductance conventional planar and 3D via-based spiral inductors, as reported in Table 6.9. Moreover, the lumped model is more accurate, as it takes into account skin effect and depletion region effects.

Table 6.10 The main advantages and disadvantages of different wireless technologies

	Far field: RF	Near field	Optical
Description	Communication range > $\frac{\lambda}{10}$	Communication range < $\frac{\lambda}{10}$	Optical wave not EM one
Pros	Long communication range Modulation is needed	Small area Magnetic field changes is proportion to $\frac{1}{R_3}$, so crosstalk is very little	Immune to interference Longer communication range
Cons	Large area, where the dimension of the antenna is larger than transistor circuits. Magnetic field changes is proportion to $\frac{1}{R}$, so crosstalk is significant	Small communication range	High cost Require optical-electrical and electrical-optical conversion, which are hard to be implemented in silicon systems
Figure of merit	Antenna gain	Coupling coefficient	Optical gain

6.4 Introduction: TSV-Based On-Chip Wireless Communications

The fast performance progress in processors and memories requires further improvement in the communication between them. However, this interface is not catching up with the processing speed. The available interface bandwidth is lower than the one predicted by Rent's rule [22, 23]. To fill this gap, wide-band, low power, and low-cost communication structures are needed. Three communication methods are considered suitable for wireless communication: RF, near field, and optical [24]. The main advantages and disadvantages of those technologies are summarized in Table 6.10.

Unlike RF and optics, near-field communication does not use wave propagation. Instead, it transfers energy through a quasi-static EM field. This enables the communication circuits to consume very low power, and eliminates wave propagation effects like fading, reflection, and phase shifting. If the distance between the transceivers is very short, near field may be more efficient than RF and optics communication.

Two main near-field methods are available: inductive and capacitive coupling. Inductive coupling has several advantages over capacitive coupling mainly because of its larger number of design options, even though the design and characterization of on-chip inductors are complicated [25, 26]. The main advantages and disadvantages of inductive coupling over capacitive coupling are summarized in Table 6.11, [27, 28]. This comparison shows that inductive-coupling communication provides the best trade-off between performance, cost, and communication distance.

Table 6.11 The main advantages and disadvantages of inductive coupling over capacitive coupling

Advantages of near-field inductive coupling over near-field capacitive coupling
Inductive coupling is driven by current change, which is not limited by supply voltage and is easily provided by current-generating circuits
Almost no high-permeability material is used in standard CMOS processes, so only relatively insignificant eddy currents generated by high-conductivity material affects the coupling magnetic field
The number of turns, which can be increased without significant additional area consumption
Longer communication distance
Disadvantages of near-field inductive coupling over near-field capacitive coupling
Consumes more power
Harder in design

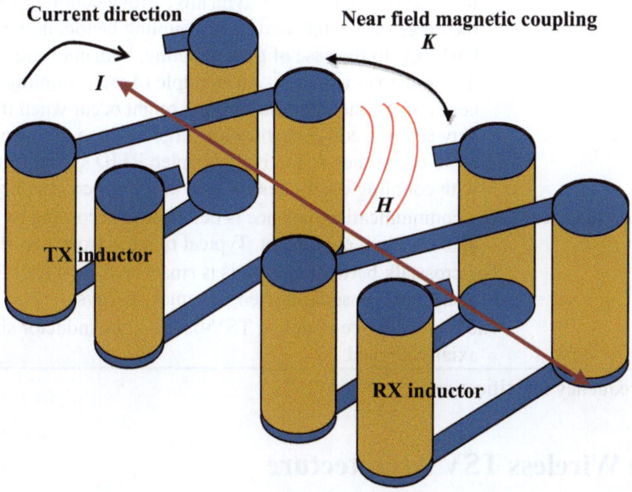

Fig. 6.12 Conceptual schematic illustrations of the wireless TSV communication on the same wafer

Inductive-coupling communication has been reported for wireless communication in the vertical direction to convey data between different tiers [29], but none considered the horizontal direction (on the same tier).

A novel on-chip wireless communication technique is introduced based on TSV spiral inductors [30, 31], as shown in Fig. 6.12. The TSV inductors exhibit near-field wireless communications, where the inductive coupling can be used for on-chip communications between adjacent blocks. The important benchmarks for this TSV-based novel wireless transmitter are summarized in Table 6.12 [32, 33].

This chapter is organized as follows. In Sect. 6.5, the architecture for the wireless TSV is introduced and simulation results are discussed. A closed-form expression for the coupling coefficient is also given. In Sect. 6.6, the closed-form equations are verified versus EM simulations. Conclusions are given in Sect. 6.7.

Table 6.12 Key operational parameters for the wireless TSV-based inductive coupling system (benchmarks)

Key operational parameters for the wireless communication system	Description
The coupling coefficient (figure of merit)	To be able to measure the efficiency of the inductive coupling between two magnetically linked conductor coils the coupling coefficient k is introduced where the transmitting and receiving coils share some magnetic flux, and it can be calculated as: $$k = \frac{M}{\sqrt{L_t L_r}} \quad (6.10)$$ where, M is the mutual inductance, L_t is transmitter self inductance, L_r is receiver self inductance, t stands for transmitter, r stands for receiver, and $0<k<1$, typically; strong coupling factors in the range of 0.3–0.6, and weak coupling factors in the range of 0.01–0.1. In the case of total coupling, both coils are subject to the same magnetic flux. An example of total coupling is a ferrite core transformer. Full decoupling might occur when the distance between two coils becomes too large or when they are perpendicular to each other. Inductively coupled RFID systems may operate with coupling coefficients as low as a few percent [32].
Communication distance	The communication distance is determined according to the acceptable coupling coefficient. Typical range is from 1 to 10 mm
Crosstalk	The crosstalk between channels is small because it uses a magnetic near field whose field strength rapidly decays
Localization	The two inductive-coupling TSV-based spiral inductor should be axially aligned

RFID radio frequency identification

6.5 The Wireless TSV Architecture

In this section, the wireless TSV architecture is introduced. The EM simulation results are presented, and a closed-form expression for the coupling coefficient is deduced.

6.5.1 The Wireless TSV: Architecture and Analysis

As stated earlier, an inductive coupling interface that uses the magnetic near field induced by TSV-based spiral inductor is introduced to permit a wireless communication channel between neighboring blocks. The turns of the two spiral inductors should be axially aligned, as the excitation currents flowing in the turns of the spiral generate a magnetic field that linked to the other coil. Referring to Fig. 6.4, the concept of the winding of the TSV-based inductor is shown, where each turn is modeled as vertical TSV. The physical parameters of the coupled TSV inductors and their typical ranges are defined in Fig. 6.15.

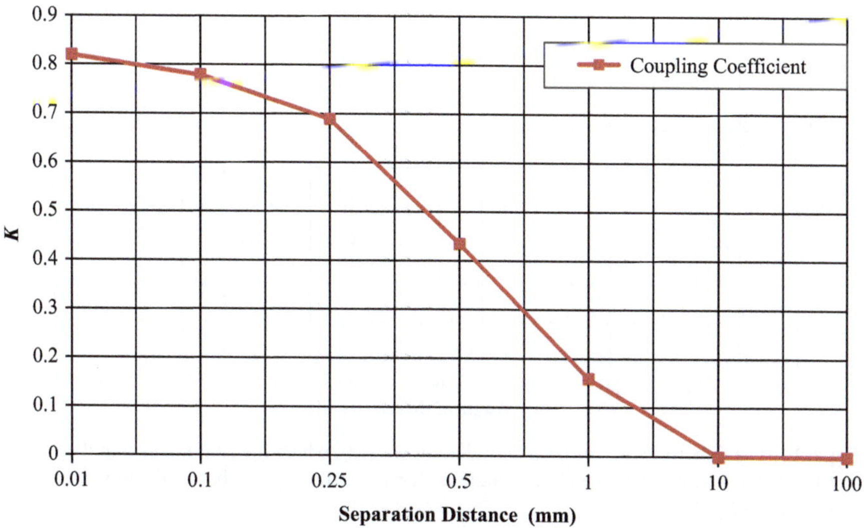

Fig. 6.13 The simulation results for the coupling coefficient *w.r.t* distance at *N* distance at n results for the coupling coefficient g axia *k* is independent of the number of turns in each coil, but the coupling decreases with distance)

6.5.2 EM Simulation Results and Discussions

The structure for the wireless TSV communication system is simulated, and the resultant *S* parameter matrix is used to characterize the operation and efficiency of the wireless TSV communication system. The inductor provides high inductance with high-quality factor and low parasitic losses, where our inductors have Qs in the 8–21 range. The high-quality factor is helpful for high-efficiency coupling.

The simulation results for the coupling coefficient *w.r.t* distance are shown in Fig. 6.13 for different *N*. Moreover, since *M* and *L* are a function of *N, K* is independent of the number of turns in each coil [34]. Thus, *K* is determined only by the gross geometry of each coil and the distance between the coils. However, it is important to increase the *N* to increase *M,* which results in increasing the receiver induced current, where, the received voltage signal can be found as [35]:

$$v_r = M\frac{\delta I_t}{\delta T} = K\sqrt{L_t L_r}\frac{\delta I_t}{\delta T} = KL\frac{I_D}{t_r} \tag{6.11}$$

where t_r is the rise time of the current signal. Therefore, a large inductor will be needed for longer communication distances. Moreover, it is recommended to use a serializer and a deserializer with this architecture to mitigate the area overhead. Closed-form equations for *K, L,* and *M* will be derived in Sect. II-C.

The directivity of this system is illustrated in Fig. 6.14, where the communication distance is fixed at 1 mm and only the angle is changed. When the inductors

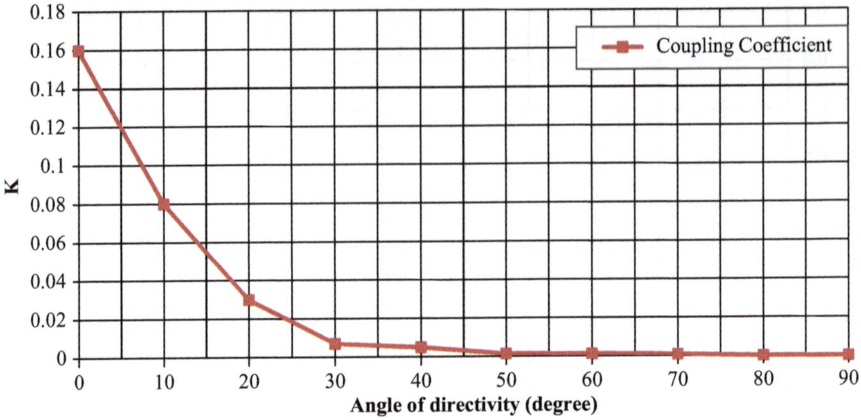

Fig. 6.14 The directivity of this system (radiation pattern), where the communication distance is fixed at 1 mm and we change only the angle in degree, assuming that both inductors are axially aligned at angle 0°

are at the same orientation (angle is 0°), coupling reaches maximum, and when the angel is 90° coupling is zero. This shows the importance of being axially aligned. Therefore, this communication is a directed communication system not isotropic one.

6.5.3 Closed-Form Expressions

Using 3D EM simulations along with curve-fitting techniques, the closed-form equation for L is given by:

$$L_{tsv\_spiral} = \frac{\mu_0 N^2 \, l_{tsv} l_m}{2\pi\,(ND_{tsv} + 0.1(N-1)\,p_m)}. \tag{6.12}$$

From the above equation, the inductance is a function of number of turns, metal track length, TSV diameter, TSV length, and spacing between tracks. Moreover, the closed-form equations for K are found using curve-fitting to EM simulations and to be given by (assuming that the inductors are at the same orientation, i.e., angle of directivity is 0°):

$$K = \frac{M}{\sqrt{L_t L_r}} = 0.85\left(\frac{80}{80 + \frac{s_{tsv}^{\;2}}{\sqrt{(l_{tsv} l_m)_t \;\; (l_{tsv} l_m)_r}}}\right)^{1.5} \tag{6.13}$$

The analytical values obtained from Eq. (6.4) and the values obtained from the simulations are correlated and their values are shown in Table 6.14. The layout area efficiency of the link for typical design values previously mentioned in Table 6.13 are (30 μm)^2. The verification of those closed-form formulae will be discussed in Sect. 6.6.

Table 6.13 TSV-based spiral inductor typical values

Design parameters	Typical range	Typical design value
Number of turns (N)	1–3	3
TSV diameter (D_{tsv})	5–100 μm	30 μm
TSV length (l_{tsv})	50–300 μm	300 μm
Metal track width (w_m)	5–100 μm	5 μm
Metal track length (l_m)	20–100 μm	20 μm
Metal track thickness (t_m)	1–5 μm	1 μm
Space between two metal tracks (P_m)	10–100 μm	10 μm
Distance between the two coils (s_{tsv})	1000–10,000 μm	1000 μm

Table 6.14 Comparison between the K values obtained from Eq. (6.4) and the values obtained from the simulations for the geometric values given in Table 6.13

Distance (mm)	K		
	Analytical value	Simulated value	Percentage error (%)
0.01	0.849	0.82	3.5
0.1	0.82	0.78	5
0.25	0.707	0.69	2.4
0.5	0.453	0.435	5
1	0.157	0.16	1.9
10	0.00028	0.00027	3.7
100	2.8×10^{-7}	2.7×10^{-7}	3.6

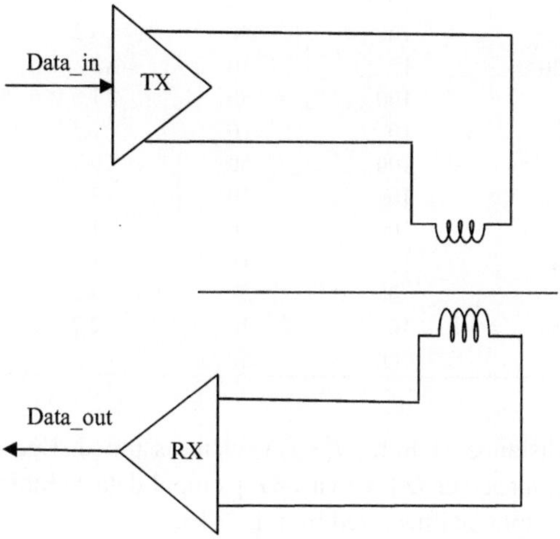

Fig. 6.15 Schematic of TSV-based inductive coupling transceiver

6.6 Wireless TSV Verification

To validate the functionality of our design, we tested our TSV-based inductive link. The verification system is very simple and consists of a transmitter and a receiver pair that uses inductive coupling. The test bed is employed in SPICE simulations (Fig. 6.15). The following parameters, L=2.3 nH, k=0.22 (communication

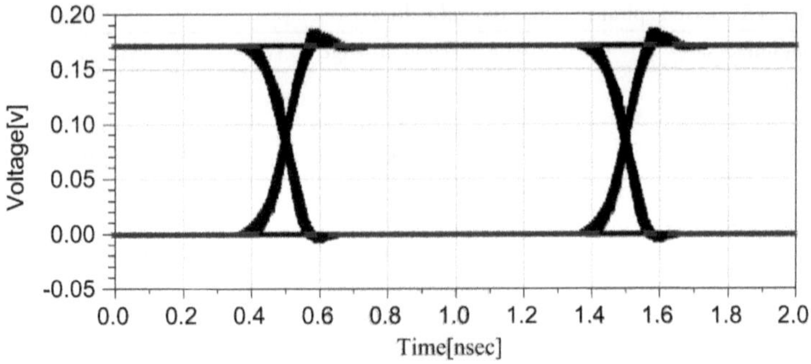

Fig. 6.16 Simulated eye diagram for output voltage of the receiver under 1-Gbps data rate. The jitter is very small. The wide eye opening indicates good data quality and clean reception

Table 6.15 Percent error between simulation and closed-form expressions of TSV-based spiral inductor inductance (l_{tsv_spiral})

D_{tsv} (μm)	l_m (μm)	p_m (μm)	l_{tsv} (μm)			
			100	150	200	300
15	10	10	2.1	2.4	3.3	4.7
	100	50	1.9	4.2	2.6	3.8
20	10	10	2.7	4.2	2	2.7
	100	50	3.9	1.4	3.2	2.5
25	10	10	1.5	1.5	3	1.3
	100	50	3.2	4.8	2.7	1.8
30	10	10	3	3.1	4.2	4
	100	50	1.5	2	4.3	3.6
35	10	10	1.5	1.5	3.1	1.3
	100	50	3.2	4.8	2.7	1.8
40	10	10	3	3.1	4.2	4
	100	50	1.5	2	4.1	3.6
45	10	10	1.5	1.5	2	1.3
	100	50	3.2	4.8	2.7	1.8
50	10	10	2.7	3.3	4.2	4
	100	50	1.5	2	4.3	2.6

distance = 1 mm), V = 1 V were assumed. Eye diagram simulation shows that the transceiver fed with a 1-Gbps input data stream can maintain very small jitter at the output as illustrated by Fig. 6.16.

To validate the closed-form equations, the approach which was applied is to simulate new random physical parameter test cases using EM simulator and compare them with the closed-form equation. For all those new random physical parameters, the S parameters of the EM simulations are correlated with the S parameters of the lumped element model whose inductance values are found using the closed-form expressions. The percent error between the simulation and the closed-form expres-

Table 6.16 Percent error between simulation and closed-form expressions of coupling coefficient (k)

l_m (μm)	l_{tsv} (μm)	S_{tsv} (mm)				
		0.1	0.3	0.7	0.7	1
20	150	1.3	2.5	2.3	1	0.7
100	300	4.8	4.3	2.6	4.8	2.3
20	150	1.3	4.3	4.1	1.9	2.2
100	300	2.4	2.6	2.2	2.6	4.4
20	150	2.6	3.3	2.1	3.4	3.6
100	300	2.1	4.3	1.4	1.7	2.9

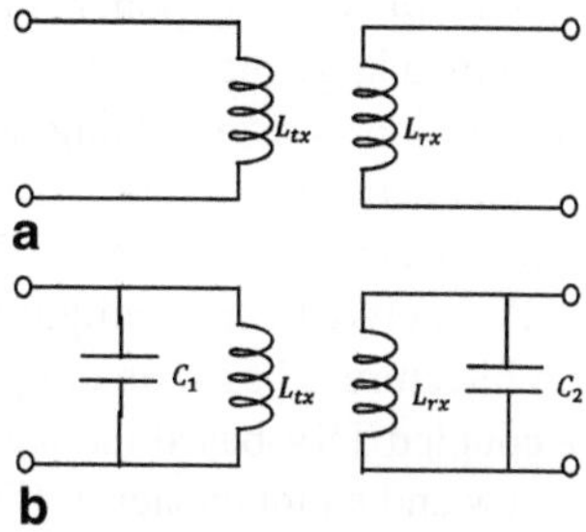

Fig. 6.17 **a** Conventional inductive coupling and **b** resonance inductive coupling

sion of TSV-based spiral inductor inductance (L_{tsv_spiral}) is shown in Table 6.15, where the maximum error is <5 %. Moreover, the percentage error between simulation and closed-form expression for the coupling coefficient (K) is shown in Table 6.16, where it is also <5 %.

6.7 Resonance Wireless Communication

A resonant inductive coupling for 3D-ICs enhances power transfer efficiency, where the resonance of an inductively coupled system increases the amount of magnetic flux linked between coils and improves the power transmission significantly [36, 37].

The concept of a resonance inductive coupling compared to a conventional one is shown in Fig. 6.17, where we have external capacitance to create a resonance at the transmitter and receiver.

To verify this model, we designed TSV-based resonance inductive power link. High-frequency structure simulation (HFSS) is used to extract the lumped element models, and these models are imported to Advanced Design System (ADS) to simulate the inductive power link. It was found that the total delivered power is larger in case of resonance inductive coupling.

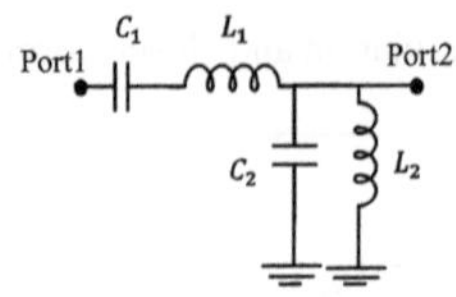

Fig. 6.18 A second-order bandpass filter, where the inductors and capacitors are all TSV-based, where, $f_0 = \frac{1}{2\pi\sqrt{L_1C_1}}$, $BW = \frac{1}{2\pi RC_2} = \frac{2\pi L_2 f_0^2}{R}$, and R is the load impedance. This filter is implemented using TSV technology

6.8 Introduction: TSV-Based Bandpass Filter

Filters are key components in the wireless systems, but these passive components occupy a large die area and suffer from lower performance [38, 39]. TSV-based 3D integration is a promising solution to improve the performance, reduce the electromagnetic interference (EMI) effect, where it reduces parasitics and losses and occupies less die area by making use of the *z* direction. This chapter introduces the TSV interconnect paradigm as a new architecture for the bandpass filter. The new architecture is based on our previously introduced TSV-based spiral inductor, where a coupled TSV-based inductor pair can be converted into bandpass filter by adding series and shunt capacitor (Fig. 6.18) [42]. TSV is also used to form the capacitors to save area (Fig. 6.19). The three regions of operation are shown in Fig. 6.20 [44].

6.9 Simulation Results

By performing EM simulation, it was found that the filter has an insertion loss of 1.5 dB at 90 GHz and 20 GHz passband from 80 to 100 GHz (Fig. 6.21). This band can be tuned by changing the dimensions of the coupled TSVs. These results show that the filter can be practically used in on-chip design. The filter size is $180 \times 120 \times 300$ μm (*xyz*). The simulation results show that TSV is a good candidate for future high performance bandpass filter designs.

6.10 Summary

In this chapter, an architecture based on TSV technology for a spiral inductor was demonstrated and characterized. Moreover, a closed-form expression for the inductance was obtained and analytical equation to calculate the maximum quality factor was derived. Simulation results indicated that our formulae are feasible. Moreover, simulations showed that our model presents better quality factor than planar spiral inductor (120% improvement) and 3D via-based spiral inductor (76% improvement) for identical inductance. Our inductors have Qs in the 8–21 range. Moreover, the SRF of the TSV-based 3D spiral inductor is 38% higher than the conventional 2D spiral inductor and 3% higher than the 3D via-based inductor. It also provides

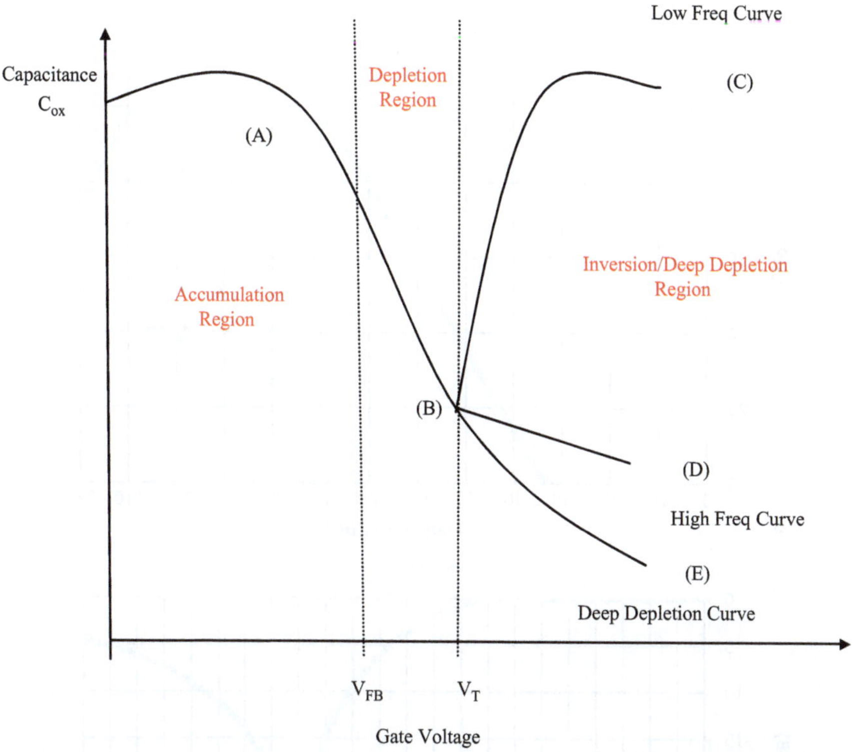

Fig. 6.19 Capacitance versus gate voltage of a TSV capacitor and its distinct operation regions: (*A*) accumulation, (*B*) depletion, and (*C*)–(*D*) inversion. Curve (*C*) is at low frequency, (*D*) at high frequency, and (*E*) at deep depletion [43]

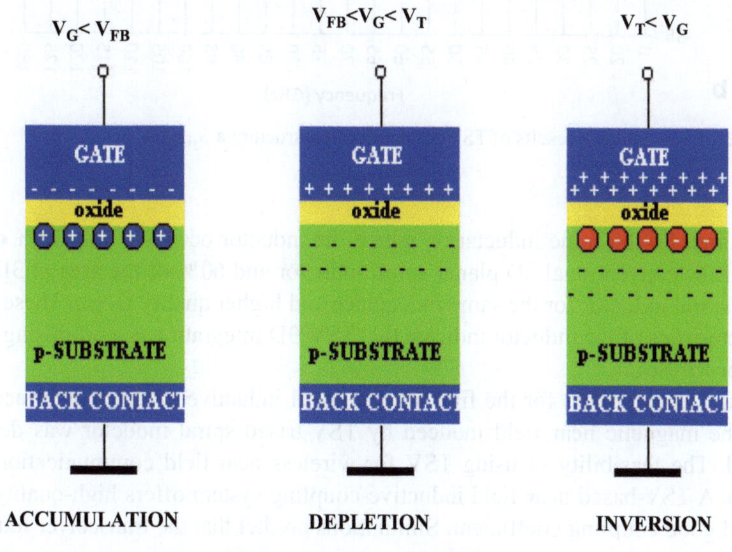

Fig. 6.20 The three regions of operation for TSV capacitance [44]

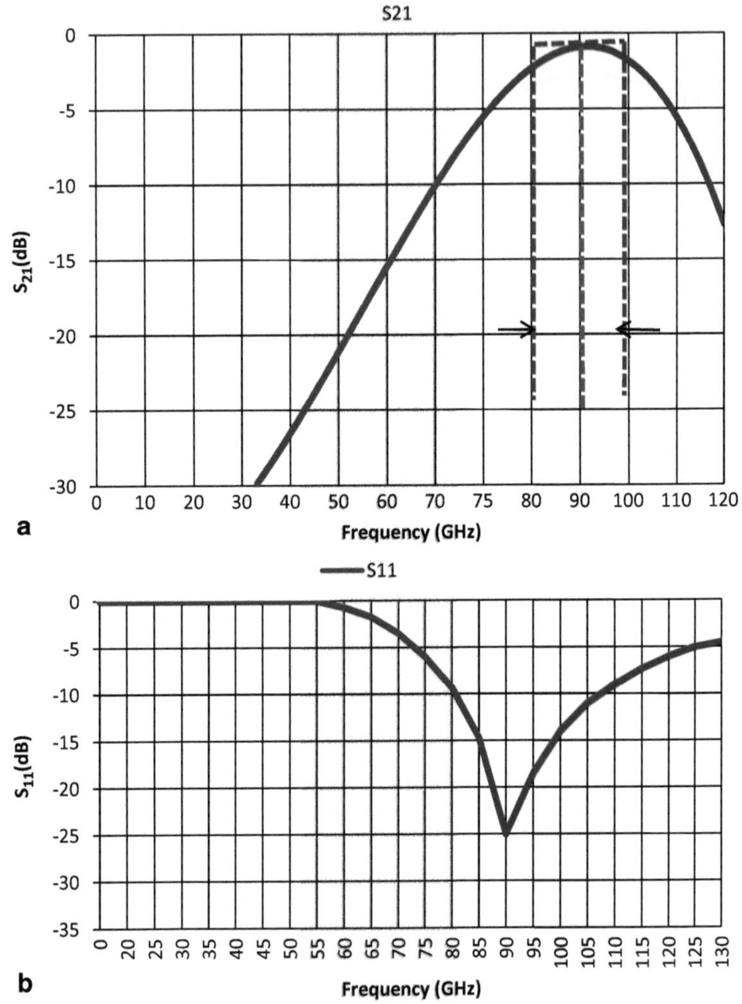

Fig. 6.21 EM-simulated results of TSV bandpass filter structure **a** S_{21} and **b** S_{11}

lower area for the same inductance, where the inductor occupies only 15 % of the area of the conventional 2D planar spiral inductor and 60 % of the area of 3D via-based spiral inductor for the same inductance and higher quality factor. These good characteristics of the inductor indicate that TSV/3D integration is a promising solution for RF-SoC.

In this chapter, and for the first time, a novel inductive coupling interface that uses the magnetic near field induced by TSV-based spiral inductor was demonstrated. The feasibility of using TSV for wireless near-field communication was shown. A TSV-based near-field inductive-coupling system offers high-quality factor and good coupling coefficient. Simulations predict that the transceiver fed with a 1-Gbps input data stream can maintain very small jitter at the output. The wireless communication link provides small area $(30\ \mu m)^2$ for typical design values. Therefore, the communication system appears to be a promising technology for wireless communication. To our knowledge, this is the first work on this topic. Moreover, the performance of TSV-based inductive coupling can be improved by using resonance method. It was found that the total delivered power is larger in case of resonance inductive coupling.

Moreover, in this chapter, a structure based on TSV technology for a bandpass filter was demonstrated and characterized. Simulations showed that the structure presents better performance than conventional ones. The TSV-based bandpass filter has an insertion loss of 1.5 dB at 90 GHz and 20 GHz passband from 80 to 100 GHz. This band can be tuned by changing the dimensions of the coupled TSVs. The bandpass filter size is $180 \times 120 \times 300\ \mu m$. These good characteristics of the bandpass filter indicate that TSV/3D integration is a promising solution for microwave applications.

References

1. J. Rosenfeld, E.G. Friedman, On-chip DC–DC converters for three-dimensional ICs. 10th international symposium on quality electronic design, IEEE 2009
2. C.P. Yue, S.S. Wong, Physical modeling of spiral inductors on silicon. IEEE Trans. Electron. Devices **47**, 560–568 (2000)
3. M. Wang, J. Li, K. D. T. Ngo, H. Xie, A novel integrated power inductor in silicon substrate for ultra-compact power supplies. 21st IEEE Applied Power Electronics Conference, APEC 2010
4. D.W. Lee, K.P. Hwang, S.X. Wang, Design and fabrication of integrated solenoid inductors with magnetic cores. 58th Electronic Components and Technology Conference (ECTC), 2008
5. A. Zolfaghari, A. Chan, B. Razavi, Stacked inductors and transformers in CMOS technology. IEEE J. Solid-State Circuits **36**(4), 620–628 (2001)
6. T.S. Horng, K.C. Peng, J.K. Jau, Y.S. Tsai, S-parameter formulation of quality factor for a spiral inductor in generalized two-port configuration. IEEE Trans. Microw. Theory Techn. **51**(11), 2197–2202 (2003)
7. M. Danesh, J.R. Long, Differentially driven symmetric microstrip inductors. IEEE Trans. Microw. Theory Techn. **50**(1), 332–341 (2002)
8. C.C. Tang, C.H. Wu, S.I. Liu, Miniature 3-D inductors in standard CMOS process. IEEE J. Solid-State Circuits **37**(4), 471–480 (2002)
9. S.S. Mohan, M.d.M. Hershenson, S.P. Boyd, T.H. Lee, Simple accurate expressions for planar spiral inductances. IEEE J. Solid-State Circuits **34**(10), 1419–1420 (1999)
10. A. Alphones, W.K. Yee, Optimization of spiral inductor on silicon. IEEE Microwave Symposium Digest, 2006
11. J.N. Burghartz, K.A. Jenkins, M. Soyuer, Multilevel spiral inductors using VLSI interconnect technology. IEEE Electron Device Lett. **17**(9), 428–430 (1996)
12. S. Kodali, D.J. Allstor, Asymmetric miniature 3D inductor. IEEE Circuits Systems, 2003
13. A. Sutono, A. Pham, J. Laskar, W.R. Smith, Development of three dimensional ceramic-based MCM inductors for hybrid RF microwave applications. IEEE Radio Frequency Integrated Circuits Symposium,1999

14. J.M. López, J. Samitier, C. Cané, P. Losantos, J. Bausells, Improvement of the quality factor of RF integrated inductors by layout optimization. IEEE Trans. Microw. Theory Techn. **48**(1), 76–83 (2000)
15. M. Ballicchia, S. Orcioni, Design and modeling of optimum quality spiral inductors with regularization and Debye approximation. IEEE Trans. Comput.-Aided Des Integr. Circuits Syst. **29**(11), 1669 (2010)
16. J. Chen, J.J. Liou, On-chip spiral inductors for RF applications: An overview. J. Semicond Technol. Sci. **4**(3), 149–167 (2004)
17. W. Yu, J.W. Bandler, Optimization of spiral inductor on silicon using space mapping. Microwave Symposium Digest, 2006
18. D.R. Pehlke, A. Burstein, M.F. Chang, Extremely high-Q tunable inductor for Si-based RF integrated circuit applications. Electron Devices Meeting, IEDM'97
19. N. Bheema, A.N. Chandorkar, 3D inductor for RF applications. Int. J. Microw. Opt. Technol. **3**(4), 179–181 (2008)
20. N.B. Rao, A.N. Chandorkar, Optically tunable spiral inductor for RF applications. The Ninth International Conference on Communications Systems, ICCS 2004
21. L.F. Tiemeijer, R.J. Havens, Y. Bouttement, H.J. Pranger, The impact of an aluminium top layer on inductors integrated in an advanced CMOS copper backend. IEEE Electron. Device Lett. **45**(11), 722–724 (2004)
22. T. Kuroda, Lower cost alternative to TSV using ThruChip interface (TCI) (Keynote). DATE 2011 Workshop on 3D Integration, March 2011
23. J. Xu, S. Mick, J. Wilson, L. Luo, K. Chandrasekar, E. Erickson, P.D. Franzon, AC coupled interconnect for dense 3-D ICs. IEEE Trans. Nucl. Sci. **51**(5), 2156–2160 (2004)
24. E. J. Marinissen, D.Y. Lee, J.P. Hayes, C. Sellathamby, B. Moore, S. Slupsky, L. Pujol, Contactless testing: possibility or pipe-dream? EDAA, 2009
25. A. More, B. Taskin, Wireless interconnects for inter-tier communication on 3D ICs. IEEE Microwave Conference (EuMC), 2010
26. E. Culurciello, A.G. Andreou, Capacitive coupling of data and power for 3D silicon-on-insulator VLSI. IEEE International Symposium on Circuits and Systems, ISCAS 2005
27. T. Zhang, R. Micheloni, G. Zhang, Z.R. Huang, J.J. Lu, 3-D data storage, power delivery, and RF/optical transceiver case studies of 3-D integration from system design perspectives. IEEE Proc **97**(1), 161–174 (2009)
28. K. Kawasaki, Y. Akiyama, K. Komori, M. Uno, H. Takeuchi, T. Itagaki, Y. Hino, Y. Kawasaki, K. Ito, A. Hajimiri, A millimeter-wave intra-connect solution. IEEE J. Solid-State Circuits **45**(12), 2655–2666 (2010)
29. S. Hu, L. Wang, Y.Z. Xiong, J. Shi, B. Zhang, D. Zhao, T.G. Lim, X. Yuan, Millimeter-wave/THz passive components design using through silicon via (TSV) technology. Electronic Components and Technology Conference, 2010
30. M. Duplessis, O. Tesson, F. Neuilly, J.R. Tenailleau, P. Descamps, Physical implementation of 3D integrated solenoids within silicon substrate for hybrid IC applications. Proceedings of the 39th European Microwave Conference, 2009
31. B. Zhang, Y.Z. Xiong, L. Wang, S. Hu, J. Shi, Y.Q. Zhuang, L.W. Li, X. Yuan, 3D TSV transformer design for DC–DC/AC–DC converter. Electronic Components and Technology Conference, 2010
32. H. Lehpamer, *RFID Design Principles*. (Artech House Inc., Norwood, 2008)
33. A. More, B. Taskin, Simulation based study of on-chip antennas for a reconfigurable hybrid 3D wireless NoC. The Proceedings of the IEEE International SoC Conference (SOCC), September 2010
34. T. Kuroda, N. Miura, Perspective of low-power and high-speed wireless inter-chip communications for SiP integration. Proceedings of the 32nd European Solid-State Circuits Conference, ESSCIRC 2006
35. R.R. Harrison, Designing efficient inductive power links for implantable devices. IEEE International Symposium on Circuits and Systems, ISCAS 2007

36. S. Han, D.D. Wentzloff, Wireless power transfer using resonant inductive coupling for 3D integrated ICs. 3DIC, 2010
37. S. Han, D. D. Wentzloff, 0.61 W/mm^2 resonant inductively coupled power transfer for 3D-ICs. CICC, 2012
38. B. Razavi, Design of millimeter-wave CMOS radios: A tutorial. IEEE Trans. Circuits Syst. Part I **56**(1), 4–16 (2009)
39. D.M. Pozar, *Microwave Engineering*. (Wiley, New York, 2005)
40. J.X. Chewn, C.Y. Cheung, Q. Xue, Integrated bandpass filter based on double-sided parallel line with an integrated conductor plane. Asian Pacific Microwave Conference, Bangkok, Thailand, December 2007
41. Jia-Sheng Hong, M.J. Lancaster, *Microstrip Filters for RF/Microwave Applications*. (Wiley, New York, 2001)
42. http://en.wikipedia.org/wiki/Constant_k_filter. Accessed 2013
43. T. Bandyopadhyay, K. Han, D. Chung, R. Chatterjee, M. Swaminathan, R. Tummala, Rigorous electrical modeling of through silicon vias (TSVs) with MOS capacitance effects. IEEE Trans. Compon. Packag. Manuf. Technol. **1**(6), 893–903 (2011)
44. A. Papanikolaou, D. Soudris, *Three Dimensional System Integration: IC Stacking Process and Design*. (Springer, New York, 2011)

Chapter 7
Imperfection in TSV Modeling

Through silicon via (TSV) is an emerging technology that enables vertical integration of silicon dies forming a single three-dimensional integrated circuit (3D-IC) stack. This chapter studies the capacitive coupling between TSVs and metal wires with wave simulation. The wave-simulation results show that coupling is not negligible when TSV is relatively short compared to the TSV width, where the aspect ratio is < 5. Therefore, TSV-to-wire capacitance needs to be considered for the computation of TSV capacitance. If the aspect ratio is > 5, the effect of metal wires is not considered. Moreover, the effect of metal lines on TSV–TSV coupling can be neglected if the pitch is less than three times the TSV diameter. Moreover, coupling between TSVs and complementary metal-oxide semiconductor (CMOS) transistors is investigated. Several isolation techniques are also introduced.

This book also studies the effects of substrate doping density on the electrical property of TSVs. With the previous lumped model for a lightly doped substrate, new lumped model for heavily doped substrate is introduced and compared with wave simulation. These lumped circuit models demonstrate acceptable modeling accuracy and may be useful for timing analysis of 3D-ICs.

7.1 Analysis of Coupling Capacitance Between TSVs and Metal Interconnects in 3D-ICs

In through silicon via (TSV)-based three-dimensional integrated circuit (3D-IC), signal integrity is becoming one of the major obstacles due to coupling, high-frequency loss, and electromagnetic (EM) radiation, where thousands of lateral and vertical interconnections are routed in a very small 3D space. In 3D-IC design, interconnects are composed of TSV and metal interconnects as depicted in Fig. 7.1. Metal interconnects of multiple layers are widely used to connect various elements on a 3D-IC with silicon interposer. Coupling capacitance between TSVs and wires surrounding the TSVs is shown in Figs. 7.2 and 7.3, where wires exist on top of TSVs in case of via-first technology (Fig. 7.2) or the TSV is surrounded by metal

K. Salah et al., *Arbitrary Modeling of TSVs for 3D Integrated Circuits*, Analog Circuits and Signal Processing, DOI 10.1007/978-3-319-07611-9_7,

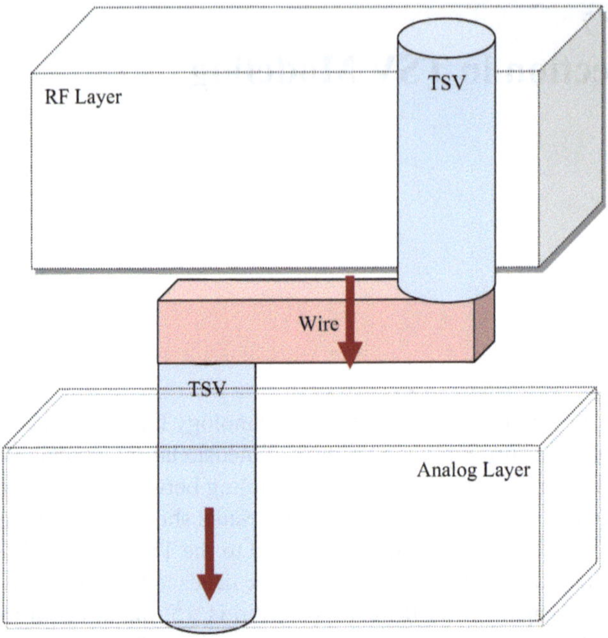

Fig. 7.1 TSV and metal lines in a 3D-IC structure

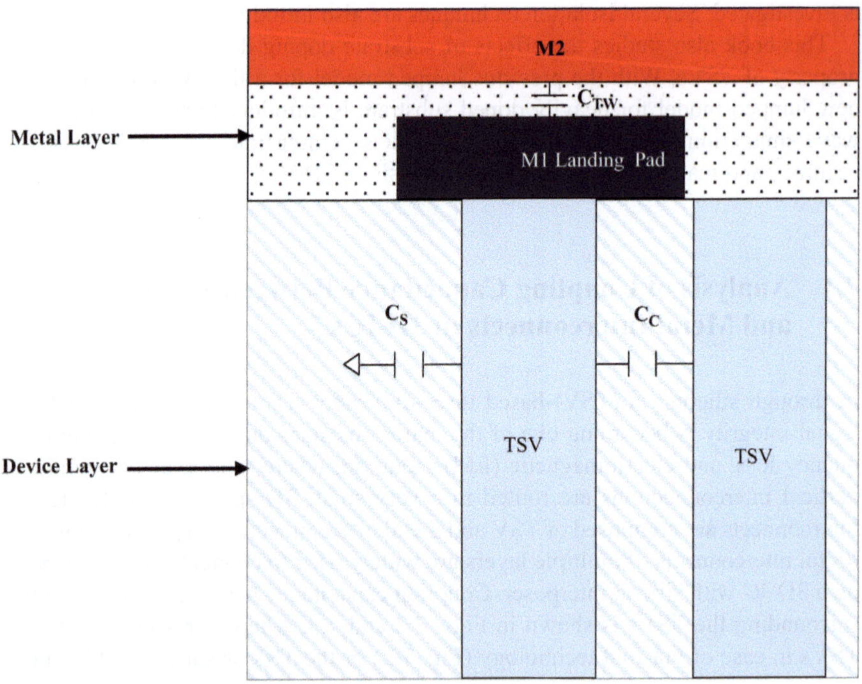

Fig. 7.2 Capacitive coupling in via-first TSV technology (vertical direction), where C_{TW} is the TSV-to-wire capacitance, C_s is the stray capacitance, and C_c is the TSV–TSV coupling capacitance

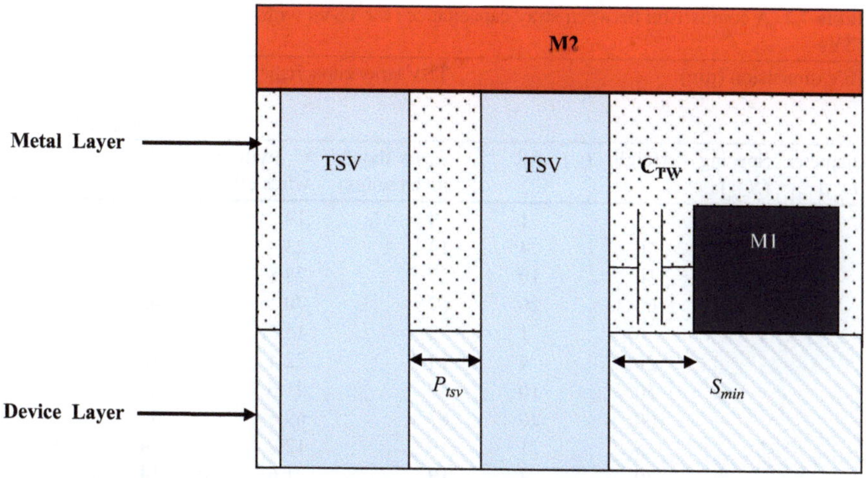

Fig. 7.3 Capacitive coupling in via-last TSV technology (lateral direction), where C_{TW} is the TSV-to-wire capacitance, C_s is the stray capacitance, and C_c is the TSV–TSV coupling capacitance

wires laterally in case of via-last technology (Fig. 7.3). TSV-to-wire coupling capacitance is affected by TSV-to-wire distance, TSV and wire dimensions, the number of surrounding TSVs and wires, and their spatial distribution.

Referring to Fig. 7.3, in via-last technology, the TSV is fabricated through all the layers from the silicon substrate to the top most metal layer; therefore, the TSV is surrounded by metal wires laterally. In via-first technology, the TSV is fabricated first and then the metals are deposited, so the TSV is surrounded by wires vertically [1]. Via-last case is a well-known case and has many closed-form solutions, but the via-first case is a new one. To analyze the effect of the coupling between a TSV and neighboring wires in metal layers, our EM simulation structure is built, where it consists of a wire above the TSV. The TSV structure is modeled with a 3D full-wave simulator (ANSYS HFSS™). The comparison of capacitances for TSVs with and without wires above the TSVs is shown in Table 7.1 for stray capacitance and in Table 7.2 for coupling capacitance.

The EM simulation results show that the capacitance between the TSV and the wire above it (C_{TW}) is not negligible when the TSV is relatively short compared to the TSV diameter, i.e., when the aspect ratio (AR) is <5. Therefore, TSV-to-wire capacitance needs to be considered for the computation of TSV capacitance. The effect of metal lines on TSV–TSV coupling (C_c), i.e, the shielding effect, and on the TSV stray capacitance (C_s) can be neglected, i.e, the error is <5 % if AR is <5 and the TSV pitch (p_{tsv}) is $<3*S_{min}$.

Table 7.1 A comparison between stray capacitances for TSVs with and without wires above the TSVs

TSV dimension (μm)				TSV stray capacitance (fF)		Error (if metal wire effects are neglected)
Diameter (D_{tsv})	Spacing (S_{min})	Height (l_{tsv})	AR	C_S (without metal wires)	C_S*(with metal wires effects)	%
5	3	5	1	9	14	55
		20	4	19	23	21
		50	10	37	38.8	5.1
		100	20	67	70	4.4
	5	5	1	9	13.5	51
		20	4	19	22.4	18
		50	10	37	41.6	4.3
		100	20	67	69.6	4
	7	5	1	9	12.9	44
		20	4	19	21.6	14
		50	10	37	38.5	4.1
		100	20	67	69.4	3.64
	10	5	1	9	12.4	38
		20	4	19	21	11
		50	10	37	38.2	3.4
		100	20	67	68.6	2.4
10	3	5	1	12.2	16	31
		20	4	23	25	8.6
		50	10	42.3	44	4
		100	20	76	78	2.6
	5	5	1	12.2	14.7	20.4
		20	4	23	24.3	5.6
		50	10	42.3	44.3	4.7
		100	20	76	77.6	2.1
	7	5	1	12.2	14	14.7
		20	4	23	24.1	5.6
		50	10	42.3	43	1.6
		100	20	76	77	1.3
	10	5	1	12.2	13.3	9
		20	4	23	23.9	3.9
		50	10	42.3	42.7	1
		100	20	76	76.8	1.05

7.2 Analysis of Coupling Capacitance Between TSVs and Active Devices in 3D-ICs

Noise coupling is a significant limitation in 3D integration, wherein some applications the input signal is in the range of μV, so it becomes very sensitive to switching noise. Therefore, understanding the impact of TSV proximity on a metal-oxide semiconductor (MOS) transistor performance is critical for integration with active circuitry without performance degradation [2–11]. TSV with MOS structure

Table 7.2 A comparison between coupling capacitances for TSVs with and without wires above the TSVs

TSV dimension (μm)				TSV coupling capacitance (fF)		Error (if metal wire effects are neglected)
Diameter (D_{tsv})	Spacing (S_{min})	Height (l_{tsv})	AR	C_C (without metal wires)	C_C^* (with metal wires effects)	%
5	3	5	1	13	20.4	57
		20	4	24	28.3	18
		50	10	42	44	4.7
		100	20	74	76.3	3.2
	5	5	1	13	20.5	58
		20	4	24	30.2	26
		50	10	42	44	4.8
		100	20	74	76.4	3.3
	7	5	1	13	18.4	42
		20	4	24	27.5	15.2
		50	10	42	44	4.7
		100	20	74	77	4.1
	10	5	1	13	16.6	28
		20	4	24	29	21
		50	10	42	44	4.6
		100	20	74	76.6	3.5
10	3	5	1	13	16.6	28
		20	4	24	26.3	9.7
		50	10	42	43.8	4.3
		100	20	74	76.2	3.1
	5	5	1	13	16.1	24.2
		20	4	24	25.5	6.3
		50	10	42	43.5	3.6
		100	20	74	76.4	3.3
	7	5	1	13	14.7	13.6
		20	4	24	25.5	6.3
		50	10	42	43.4	3.5
		100	20	74	75.6	2.2
	10	5	1	13	13.8	6.6
		20	4	24	25.15	4.8
		50	10	42	42.5	1.3
		100	20	74	74.88	1.2

is shown in Figs. 7.4 and 7.5 for bulk silicon (Si) and silicon on insulator (SOI) respectively.

3D integration can be implemented with various device technologies. Here, we focus on 65-nm bulk complementary MOS (CMOS) devices integrated with TSV. The channel length is 1 μm in order to avoid short-channel effects. The doping level is 10^{+15} cm^{-3}. The electrical characteristics of the simulated transistor are summarized in Table 7.3. The coupling network can be simplified as shown in Fig. 7.6 and the physical interpretation is shown in Table 7.4.

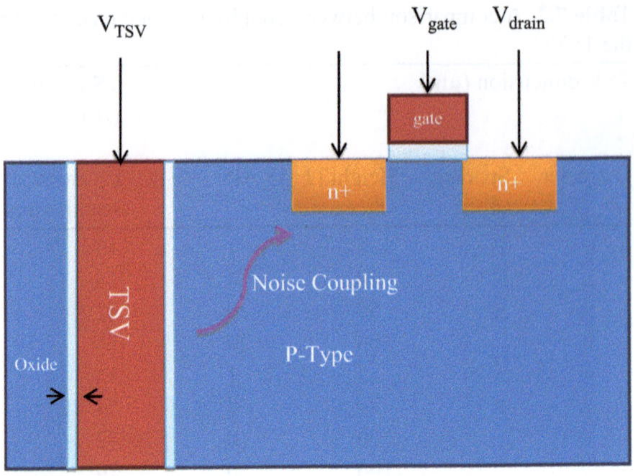

Fig. 7.4 TSV-to-MOS structure, bulk Si

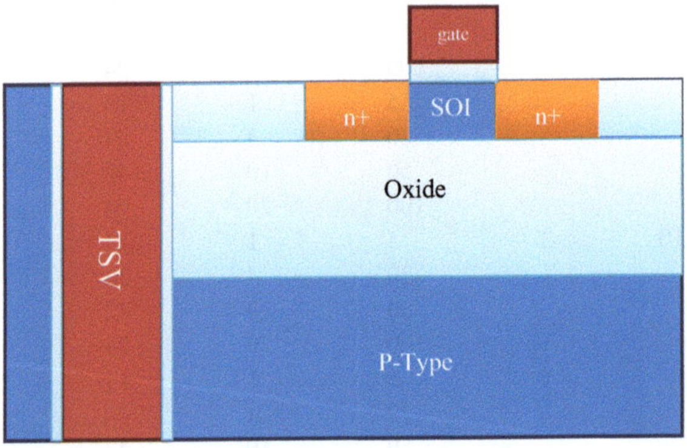

Fig. 7.5 TSV-to-MOS structure, SOI

A square voltage (f=200 MHz, 50 ps-long rise and fall ramp time, 1.2 V peak voltage) is applied on the copper (Cu) TSV to see its effect on CMOS as an aggressor digital signal. Simulation test bed is shown in Fig. 7.4, where technology computer aided design (TCAD) simulation (Silvaco ATLAS, [12]) is used and the results show that:

1. It was demonstrated that substrate spike noise is produced in the substrate by the TSV structures, when a square signal is applied.
2. A thicker *oxide* for TSV isolation reduces significantly current variations because of the decrease of TSV oxide capacitance.
3. Effect on saturation current ($V_{GS}=V_{DS}=1.2$ V) shows that there is a 6 μA/μm of CMOS transistor channel length spike variation.
4. Effect on leakage current ($V_{GS}=0$ V and $V_{DS}=1.2$ V) shows that there is a 1.2 μA/μm of CMOS transistor channel length spike variation.
5. The results demonstrate superior noise isolation for *SOI* substrates compared to bulk silicon due to the buried oxide layer capacitance.
6. Drain saturation current slightly depended on the distance from a TSV to an MOS transistor.

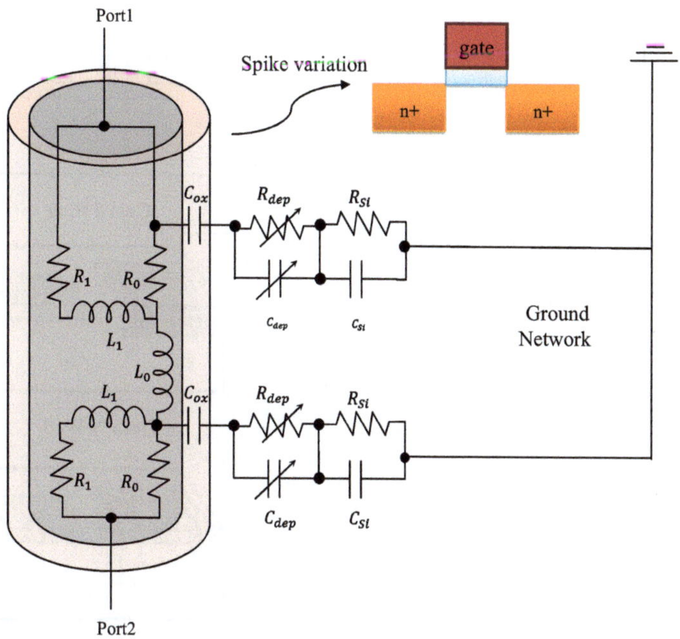

Fig. 7.6 Circuit model to represent noise coupling between TSV and MOS. The model is composed of R_0, L_0, R_1, L_1, C_{ox}, C_{si}, C_{dep}, R_{si}, and R_{dep}

Table 7.3 Electrical characteristics of the simulated NMOS transistor with a channel length of 1 µm

Property	Symbol
Threshold voltage	V_T
Saturation drain current at $V_{GS}=V_{DS}$	I_{DSAT}
Leakage current at $V_{GS}=0$ V	I_{Leak}

Table 7.4 The model lumped elements and its physical interpretation

Circuit element	Physical meaning
R_0, L_0	Inductance and ohmic loss of the conductor
R_1, L_1	Skin effect of the conductor
C_{ox}	Capacitance of the oxide
C_{dep}, R_{dep}	Silicon substrate depletion region capacitance and resistance
C_{si}, R_{si}	Silicon substrate capacitance and resistance

7. It seems that the TSV has quite no influence on *digital circuit* behavior (such as inverter).
8. For *analog circuits,* the impact of the TSV depends on the frequency operation and the noise figure needs.
9. The TSV parasitic capacitances and the substrate impedance acts as a *differentiator* for the dynamic signal applied on the TSV electrode.

Fig. 7.7 Noise isolation techniques [11]

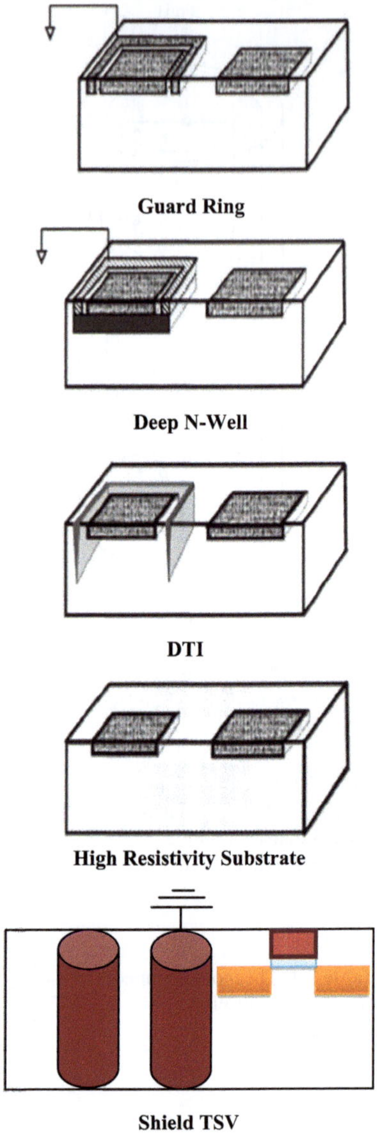

10. As the TSV *length* decreases, the capacitance decreases and the noise peak also decreases.

The impact of TSV must be annihilated by using a guard ring, deep N-well, deep trench isolation (DTI), high-resistivity substrate, or shield TSV (Fig. 7.7) [11].

1. The *guard ring* is the most popular solution for the noise coupling; its dimension should be the same as the victim device and it is connected to a ground network.

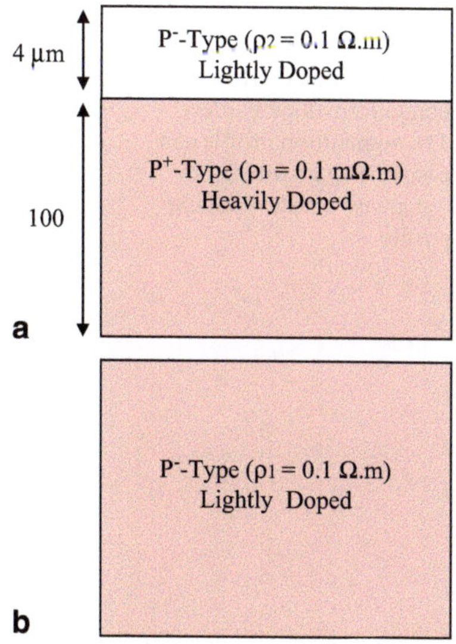

Fig. 7.8 **a** Heavily doped substrate and **b** lightly doped substrate

The reduction of noise is independent of the frequency. It absorbs the substrate noise current. The greater the distance from the victim the better the isolation.

2. The *deep N-well* is placed under the circuit block and is also connected to the ground network.
3. *DTI* is made from high-resistivity polysilicon.
4. Also, increasing the *resistivity* of the substrate attenuates the noise signals.
5. For the *shield TSV*, it is placed near the aggressor TSV and connected to the ground network.

7.3 Effect of Nonuniform Substrate-Doping Profile on the Electrical Performance of a TSV

In conventional bulk CMOS technologies, there are mainly two types of substrates that are used [13]. The first substrate type consists of a thick layer of heavily positively doped silicon (p^+) with a thin lightly positively doped silicon (p^-) on top as illustrated in Fig. 7.8. This kind of technology is referred to as epitaxial substrate or heavily doped substrate (e.g., $\sigma = 0.1 \sim 100$ kS/m, $\rho = 0.01 \sim 10$ mΩ.m). The thickness of the epitaxial layer is typically about 4 µm, where the thickness of a chip is commonly from 100 to 300 µm. In modeling, the layer of the heavily doped silicon may be approximated by a single node due to its high conductance [14].

Therefore, the substrate noise in heavily doped substrates tends to be uniform over the chip area. Historically, heavily doped substrates have been widely used due to the low risk of latch-up [15]. The second substrate type is the uniform lightly

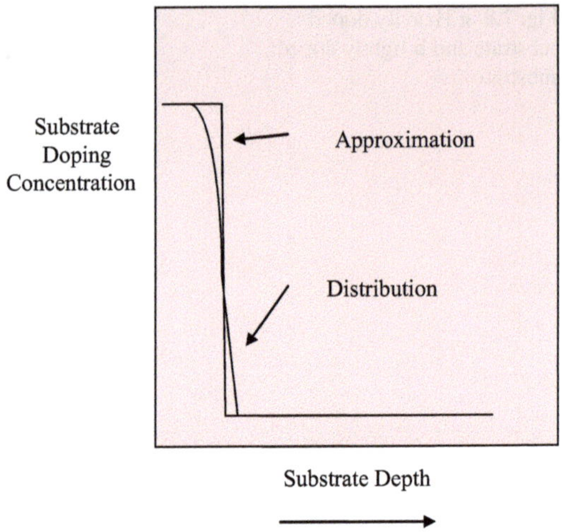

Fig. 7.9 Actual substrate doping distribution and its approximation (vertical nonuniform doping effect). The nonuniform profile can easily be approximated by a stepwise constant doping profile

doped substrate, which is illustrated in Fig. 7.8. In this substrate, the resistivity is higher than in the heavily doped substrate (e.g., $\sigma = 1 \sim 100$ S/m, $\rho = 0.01 \sim 1$ Ω.m). Therefore, the substrate coupling is smaller in the lightly doped substrate. For this reason, the lightly doped substrate is considered to be a better choice for mixed-signal ICs than the heavily doped substrate [16, 17]. In this section, the effects of nonuniform doping profile on the TSV model accuracy are studied, two lumped element model for the TSV in a heavily doped substrate, and lightly doped one are introduced. Those two models are combined into one unique model. The model is verified using a 3D field solver.

Considering the current 3D TSV technology and near-future 3D applications, a circular Cu TSV 10 μm in diameter and 100 μm in height is selected for this work as the baseline TSV structure in a two-stratum configuration. Cu is selected as the TSV conductive material based on the present TSV technology. A thin layer of 0.3 m SiO_2 serves as the isolation layer between the Si substrate and TSV conductive material.

In Chap. 2, a lumped model for the TSV was introduced under the assumptions of a uniform doping. Here, the effects of nonuniform substrate doping profile on the electrical performance of a TSV are studied. The nonuniform profile will be modeled as multilayer substrate with different conductivities as shown in Fig. 7.9, i.e., the nonuniform profile can easily be approximated by a stepwise constant doping profile [18]. So, the TSV penetrates downward two stratified layers of silicon (heavily doped and lightly doped), as shown in Fig. 7.10.

The TSV structure is modeled with a 3D full-wave simulator. The corresponding lumped element model for the lightly doped substrate is shown in Fig. 7.11, and its element interpretation is summarized in Table 7.4. Since TSVs are made of conducting material such as Cu or tungsten, a TSV is modeled as a series connection of a resistor and an inductor. The silicon dioxide insulator between TSV and silicon

Fig. 7.10 Multilayer substrate (stratified substrate approximation)

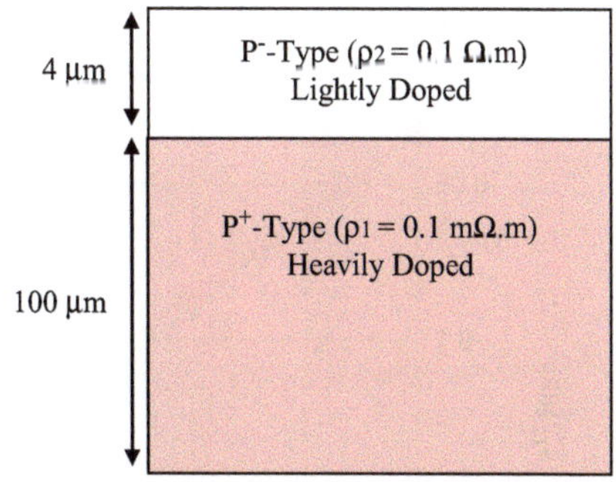

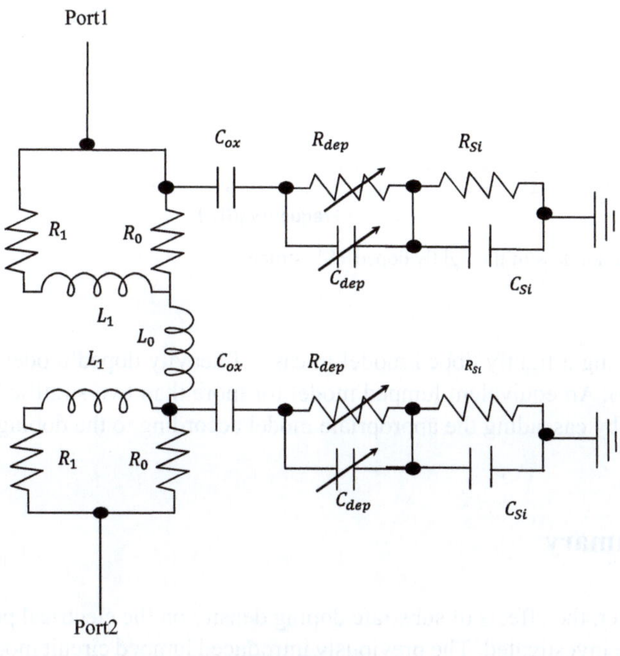

Fig. 7.11 Lumped circuit model for lightly doped case

substrate is modeled as a capacitor. On the other hand, the silicon substrate can be modeled as a capacitor in parallel with a resistor.

The lumped circuit model is simulated for lightly doped substrate and the *S* parameter results are compared with the 3D full-wave simulation as shown in Fig. 7.12, where they show good agreement up to 10 GHz. For the heavily doped substrate, the lumped circuit model is introduced in Fig. 7.13, where the heavily doped silicon can be approximated to a single node due to its high conductance. The model's element interpretations are summarized in Table 7.5. The lumped circuit model is simulated and the *S* parameter results are compared with the 3D full-wave simulation as shown in Fig. 7.14, where they show good agreement. Simulations

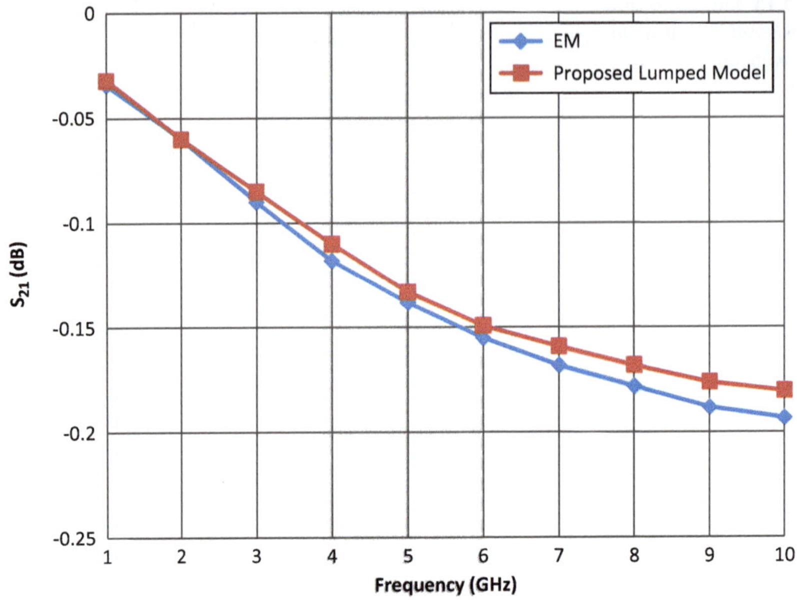

Fig. 7.12 Insertion loss of the lightly doped TSV structure

show that using a lightly doped model in case of heavily doped model exhibits an error of 18 %. An equivalent lumped model for more than two stratified layers can be obtained by cascading the appropriate model according to the doping level.

7.4 Summary

In this chapter, the effects of substrate doping density on the electrical performance of TSV were investigated. The previously introduced lumped circuit model for TSV structure was used for a lightly doped silicon structure. A new lumped circuit model for a heavily doped silicon substrate was introduced and its physical understanding was explained. Both circuit models for the lightly doped and heavily doped cases were validated using full-wave simulations up to 10 GHz. An equivalent lumped model for more than two stratified layers can be obtained by cascading the appropriate model according to the substrate type. Moreover, parasitic coupling capacitive between TSVs and surrounding wires was analyzed and studied, where our model considers coupling with surrounding wires in vertical directions. The simulation results for the electrical characteristics of the coupling between TSVs and metal lines in 3D-ICs showed that the coupling is not negligible when TSV is relatively short compared to the TSV width, where the aspect ratio is <5. Therefore, TSV-to-wire capacitance needs to be considered for the computation of TSV capacitance.

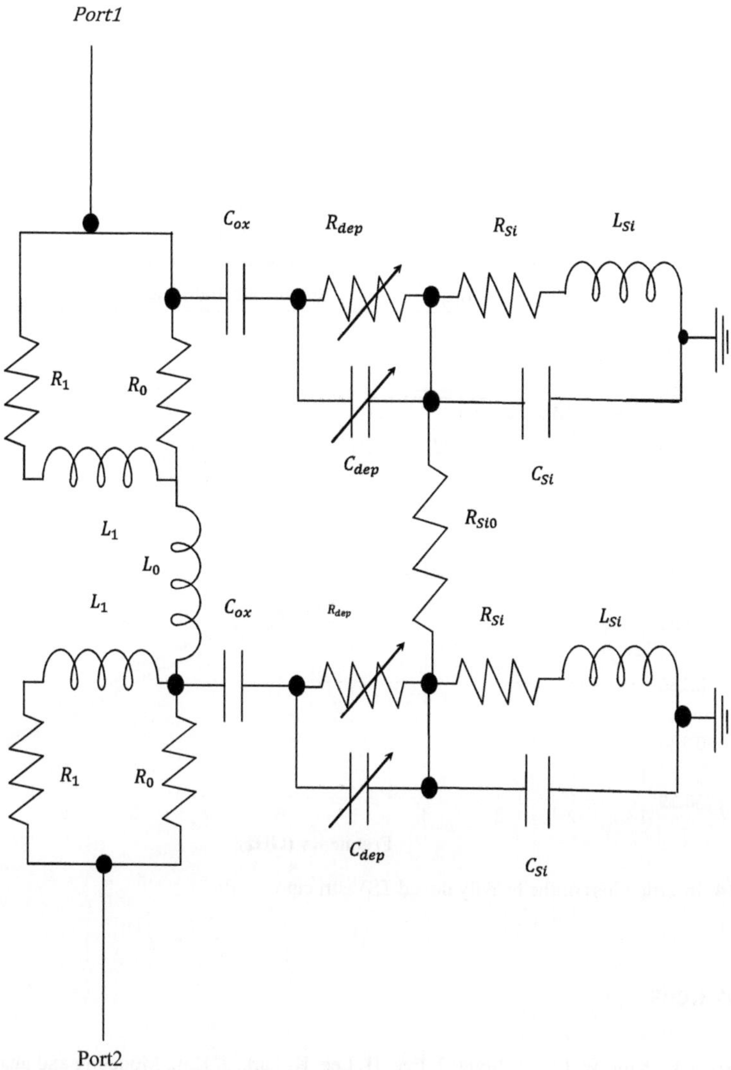

Fig. 7.13 Lumped circuit model for heavily doped case

But, if the aspect ratio is larger than 5, the effect of metal wires is not considered. Moreover, the effect of metal lines on TSV–TSV coupling can be neglected if the pitch is less than three times the TSV diameter. This analysis is very helpful for timing analysis in 3D-ICs. Moreover, in this chapter, noise coupling between TSVs and CMOS was investigated. Several isolation techniques were also introduced.

Table 7.5 The heavily doped model lumped elements and its physical interpretation

Circuit element	Physical meaning
R_0, L_0	Inductance and ohmic loss of the conductor
R_1, L_1	Skin effect of the conductor
C_{ox}	Capacitance of the oxide
C_{dep}, R_{dep}	Silicon substrate depletion region capacitance and resistance
C_{si}, R_{si}	Silicon substrate capacitance and resistance
L_{si}	The self-partial inductance associated with the silicon substrate
R_{si0}	Silicon substrate lateral resistance

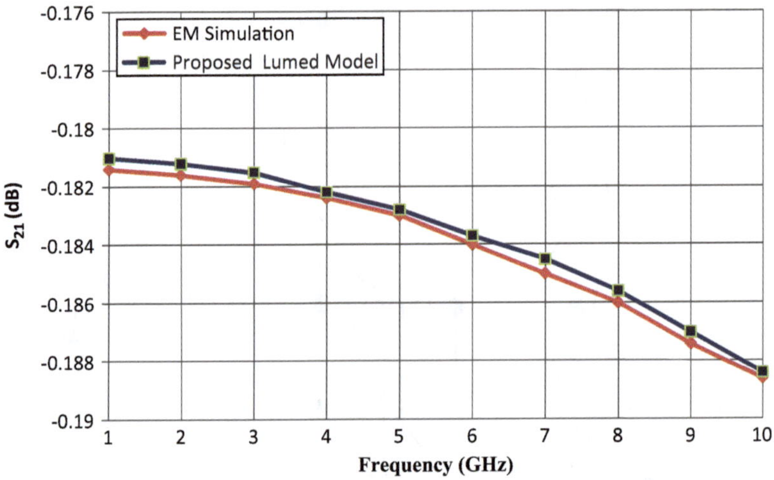

Fig. 7.14 Insertion loss of the heavily doped TSV structure

References

1. K. Yoon, G. Kim, W. Lee, T. Song, J. Lee, H. Lee, K. Park, J. Kim, Modeling and analysis of coupling between TSVs, metal, and RDL interconnects in TSV-based 3D IC with silicon interposer. Electronics Packaging Technology Conference,11th, pp. 702–706, IEEE, Singapore, 9–11 Dec 2009
2. M. Rousseau, M-A. Jaud, P. Leduc, A. Farcy, A. Marty, Impact of substrate coupling induced by 3D-IC architecture on advanced CMOS technology. European Microelectronics and Packaging Conference, IEEE, Rimini, Italy, 15–18 June 2009
3. H. Chaabouni, M. Rousseau, P. Leduc, A. Farcy, R. El-Farhane, A. Thuaire, G. Haury, A. Valentian, G. Billiot, M. Assous, F. De Crecy, J. Cluzel, A. Toffoli, D. Bouchu, L. Cadix, T. Lacrevaz, P. Ancey, N. Sillon, B. Flechet, Investigation on TSV impact on 65 nm CMOS devices and circuits. Electron Devices Meeting (IEDM), IEEE, San Francisco, 6–8 Dec 2010
4. N.H. Khan, S.M. Alam, S. Hassoun, Through-Silicon Via (TSV)-induced noise characterization and noise mitigation using coaxial TSVs. Conference on 3D System Integration, IEEE, San Francisco, 28–30 Sept 2009

5. M. Rousseau, M-A. Jaud, P. Leduc, A. Farcy, A. Marty, Through-silicon via based 3D IC technology: Electrostatic simulations for design methodology. IMAPS, Device Packaging Conference, Phoenix, 2008
6. P-J. Yang, P-T. Huang, W. Hwang, Substrate noise suppression technique for power integrity of TSV 3D integration. Circuits and Systems (ISCAS), 2012 IEEE International Symposium, Seoul, 20–23 May 2012
7. J. Cho, J. Shim, E. Song, J.S. Pak, J. Lee, H. Lee, K. Park, J. Kim, Active circuit to through silicon via (TSV) noise coupling. Electrical Performance of Electronic Packaging and Systems, IEEE, Portland, 19–21 Oct 2009
8. C. Xu, R. Suaya, K. Banerjee, Compact modeling and analysis of through-si-via-induced electrical noise coupling in three-dimensional ICs. IEEE Transactions on Electron Devices, Vol. 58, No. 11, Nov 2011
9. S. Uemura, Y. Hiraoka, T. Kai, S. Dosho, Isolation techniques against substrate noise coupling utilizing through silicon via (TSV) process for RF/mixed-signal SoCs. IEEE J. Solid-State Circuits **47**(4) (April 2012)
10. N.H. Khan, S.M. Alam, S. Hassoun, Mitigating TSV-induced substrate noise in 3-D ICs using GND Plugs. 12th International Symposium on Quality Electronic Design (ISQED), IEEE, Santa Clara, 14–16 March 2011
11. E. Salman, Noise coupling due to through silicon vias (TSVs) in 3-D integrated circuits. IEEE International Symposium on Circuits and Systems, Rio de Janeiro, 15–18 May 2011
12. Silvaco International (Atlas User's Manual, USA, 2008), www.silvaco.com. Accessed 2012
13. G. Veronis, Y.C. Lu, R.W Dutton, Modeling of wave behavior of substrate noise coupling for mixed-signal IC design. Proceedings of the International Symposium on Quality Electronic Design, pp. 303–308, Stanford University, CA, 2004
14. D.K. Su, M.J. Loinaz, S. Masui, B.A. Wooley, Experimental results and modeling techniques for substrate noise in mixed-signal integrated circuits. IEEE J. Solid-State Circuits **28**, 420–430 (1993)
15. T. Blalack, J. Lau, F.J.R. Clément, B.A. Wooley, Experimental results and modeling of noise coupling in a lightly doped substrate. International Electron Devices Meeting, IEDM '96 Tech. Dig., pp. 623–626, San Fransisco, 8–11 Dec 1996
16. X. Aragonès, A. Rubio, Experimental comparison of substrate noise coupling using different waver types. IEEE J. Solid-State Circuits **34**, 1405–1409 (1999)
17. H. Wang, J. Kim, Y. Shi, J. Fan, The effects of substrate doping density on the electrical performance of through-silicon-via (TSV). Asia-Pacific Symposium & Exhibition on Electromagnetic Compatibility, Jeju Island, 16–19 May 2011
18. BSIM3v3 Manual Copyright © 1995, 1996, UC Berkeley

Chapter 8
New Trends in TSV

In this chapter, several new directions are discussed for through silicon via (TSV) fabrication and use in design. Performance comparison between air-Gap gap-based coaxial TSV and conventional circular TSV is presented. The comparison shows that the air-gap TSVs reduce the overall parasitic capacitance and the overall energy loss compared to the conventional circular TSV or conventional coaxial TSV. High-performance 3D TSV interconnects are important for reliability. Choice of the filler material is also a critical issue as thermal incompatibility, electromigration and, high resistivities are still a bottleneck. In this book, SW-CNT bundles are investigated as a prospective filler material for TSV compared to conventional filler materials like Cu, W, and polysilicon. It is found that SW-CNT bundles exhibit unique electrical, thermal, and mechanical characteristics that can be used to fabricate better TSV interconnects.

Moreover, in this chapter, TSV-based adiabatic logic based on the adiabatic switching principle is presented and analyzed. Adiabatic logic is a design technique for minimizing the energy dissipation. Its major limitation is the requirement for passive components, which cannot be efficiently integrated into current generation integrated circuits (ICs). TSV-based 3D heterogeneous integration may enable efficient integration of these passive elements, which were not practically feasible in the past due to technology limitations.

8.1 Performance Comparison Between Air-Gap-Based Coaxial TSV and Conventional Circular TSV

A TSV can be circular or rectangular, but the stress profile of a circular through silicon via (TSV) is better than a rectangular one. Conventional circular TSV is lossy due to low-resistivity substrate and has large signal degradation at high frequency. In order to improve the radio frequency (RF) performance, coaxial TSV is presented which is able to suppress undesirable substrate loss and provide good impedance matching. Also, fanlike air-gap coaxial TSV is presented for further improvement.

K. Salah et al., *Arbitrary Modeling of TSVs for 3D Integrated Circuits*,
Analog Circuits and Signal Processing, DOI 10.1007/978-3-319-07611-9_8,

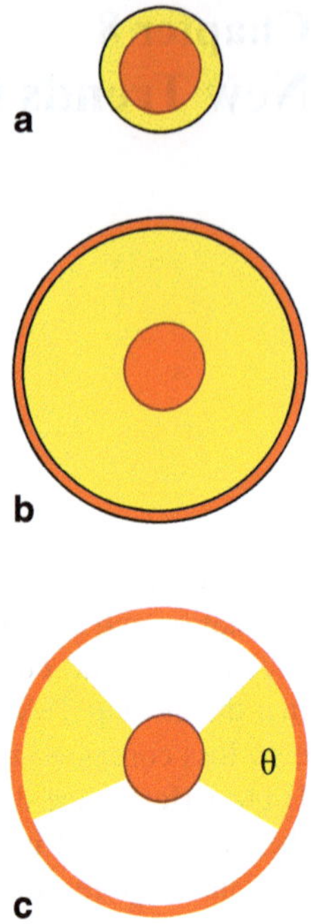

Fig. 8.1 **a** Conventional circular TSV, **b** coaxial TSV, **c** air-gap TSV

Conventional circular TSV, coaxial TSV, and air-gap TSV are depicted in Fig. 8.1 [1, 2].

Electrical modeling is performed using Ansoft's High Frequency Structure Simulation (HFSS). Electromagnetic (EM) simulations of the three types of TSV show the following results:

1. Air-gap TSV has better insertion coefficient and less reflection coefficient (Fig. 8.2).
2. If θ is decreased, the insertion coefficient increases (Fig. 8.3).
3. If we have more fans (Fig. 8.4), the insertion coefficient improves (Fig. 8.5).

However, air-gap TSV may be a challenge in fabrication. So, performance comparison between air-gap-based coaxial TSV and conventional circular TSV shows that the air-gap TSVs reduce the overall parasitic capacitance and the overall energy loss compared to the conventional circular TSV or conventional coaxial TSV.

8.2 Characterization of SW-CNT-Based TSV

High-performance 3D TSV interconnects are important for reliability, choice of the filler material is also a critical issue as thermal incompatibility, electromigration, and high resistivity are still a bottleneck. In this chapter, SW-CNT bundles as a prospective filler material for TSV are investigated compared to conventional filler materials like Cu, W, and polysilicon [3].

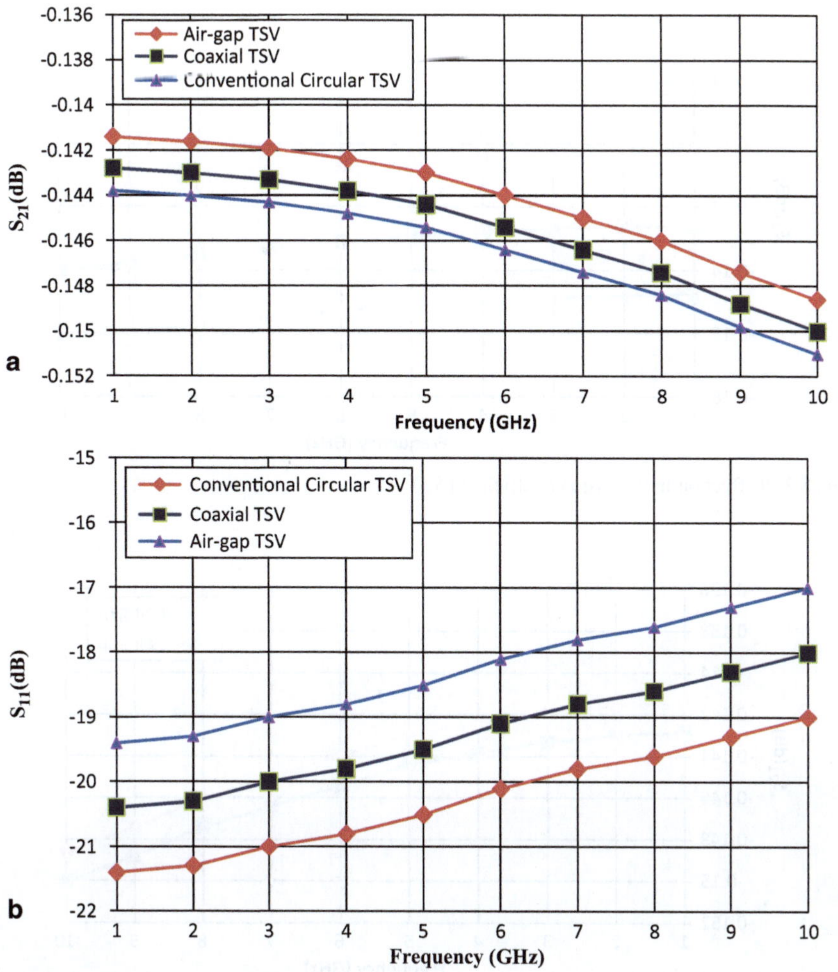

Fig. 8.2 *S* parameters comparison between conventional circular TSV, coaxial TSV, and air-gap TSV **a** S_{21}, **b** S_{11}

The desired properties of via-filling material include void-free filling, lower stress, excellent electrical current carrying capability and conductivity, and good thermal performance. SW-CNT bundles exhibit unique electrical, thermal, and mechanical characteristics that can be used to fabricate better TSV interconnects.

Carbon nanotubes (CNT) are cylinders of graphite, same as the material in a pencil. CNT can be metallic or semiconductor, it all depends on the angle of rolling and diameter of the nanotube. CNT-based TSV has excellent current carrying capability higher than copper, as it does not exhibit skin effect at high frequency (Cu 10^6 A/cm^2, CNT 10^9 A/cm^2). Moreover, it has better thermal stability than copper as it works fine at higher temperatures, thus eliminating the problem of electromigration with Cu interconnects.

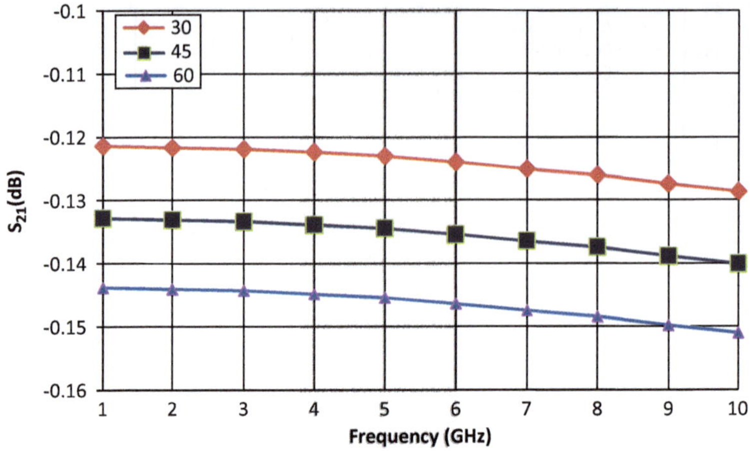

Fig. 8.3 θ effect on transmission coefficient (S_{21})

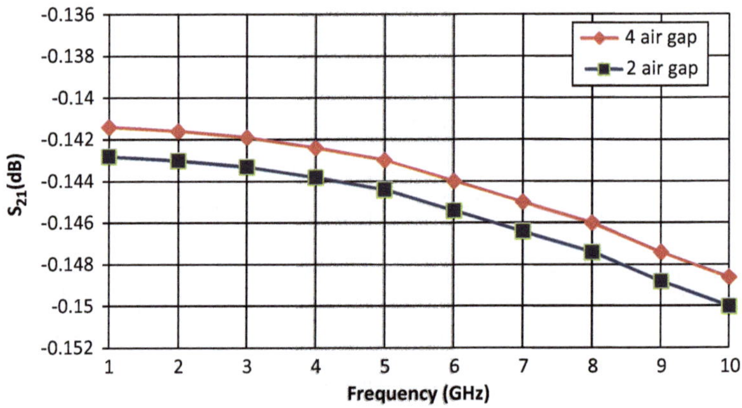

Fig. 8.4 Structure of **a** two-air-gap TSV, **b** four-air-gap TSV

Fig. 8.5 Four-air-gap and two-air-gap S_{21} comparison

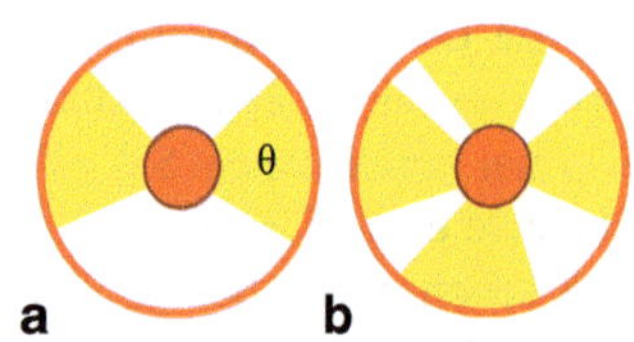

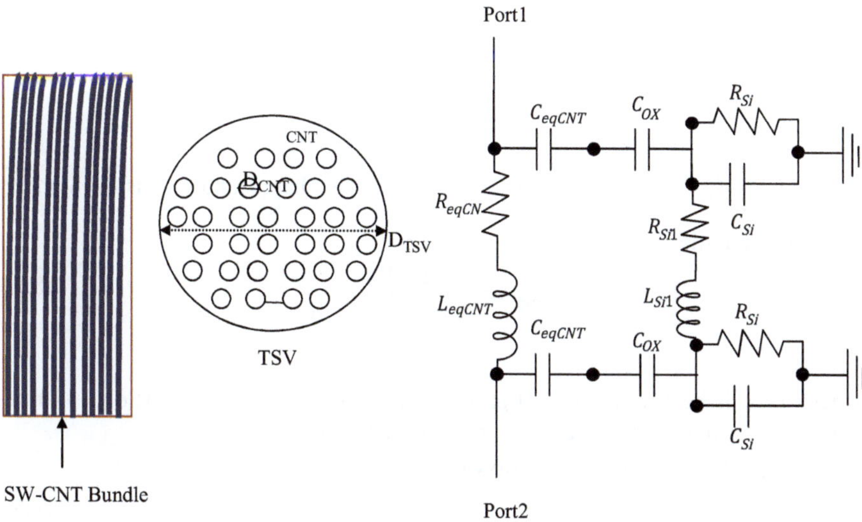

Fig. 8.6 CNT-based TSV structure and its lumped element model

CNT is a fairly new molecular structure of carbon and has been applied to the semiconductor industry recently. It has unique electrical and mechanical properties that it can be shaped to act as conductor, semiconductor, or insulator. CNT is able to reduce semiconductor manufacturing processing steps that incorporate copper, such as the forming of insulator to prevent copper from diffusing through the gate. CNTs are cylindrical structures based on the hexagonal lattice of carbon atoms that covalently bond together and form crystalline graphite. CNTs can be looked at as single molecules, due to their small size (~1 nm in diameter and ~1 μm length).

There are two major types of CNTs: single-walled carbon nanotubes (SW-CNT) and multi-walled carbon nanotubes (MW-CNT). MW-CNT can be described as an assembly of concentric single-walled carbon nanotubes with different diameters.

The electrical circuit model for CNT-based TSV takes into consideration the quantum resistance, the quantum capacitance, and the kinetic inductance in addition to conventional insulator and substrate-based parasitics, as depicted in Fig. 8.6 [4–8].

The equivalent resistance is given by:

$$R_{eqCNT} = R_{deg} + R_{scat}. \tag{8.1}$$

Spin and sub-lattice degeneracy of electrons in graphene is given by:

$$R_{deg} = \frac{h}{4e^2 n_{CNT}}. \tag{8.2}$$

Scattering effect in case of $L_{CNT} > \lambda$ is given by:

$$R_{scat} = \frac{h}{4e^2 n_{CNT}} \frac{L_{CNT}}{\lambda} \tag{8.3}$$

where

$\frac{h}{4e^2} = \frac{10^6}{155}\Omega$, h is Planck's constant, e is electron charge, and n_{CNT} is the number of CNT in bundle and is given by:

$$n_{CNT} = \frac{\pi(\frac{D_{TSV}}{2})^2}{\frac{\sqrt{3}}{2}(D_{CNT} + S_{CNT})} \tag{8.4}$$

where L_{CNT} is height of CNT, D_{CNT} is diameter of a single SW-CNT; *e.g.*, = 1 nm, λ is the mean free path of electrons in CNT = 1.6 μm.

The equivalent capacitance is given by:

$$C_{eqCNT} = C_{qnt} + C_{electro}. \tag{8.5}$$

The quantum capacitance is given by:

$$C_{qnt} = \frac{2e^2}{hv_f} \tag{8.6}$$

where v_f is the Fermi velocity of CNT = 8×10^5 m/s.

Electrostatic capacitance which is the parallel-plate capacitance of isolated SW-CNT *w.r.t* near and far SW-CNT is given by:

$$C_{electro} = 2\pi\varepsilon \left[\frac{4 + 3\sqrt{n_{CNT}}}{5 ln\left(\frac{D_{TSV}}{D_{CNT}}\right)} + \frac{\sqrt{n_{CNT}} - 2}{2 ln\left(\frac{2D_{TSV}}{D_{CNT}}\right)} \right]. \tag{8.7}$$

The equivalent inductance is given by:

$$L_{eqCNT} = L_{mag} + L_{kin}. \tag{8.8}$$

The magnetic inductance is given by:

$$L_{mag} = \frac{\mu}{2\pi}.L_{CNT}.\left[ln\left(\frac{L_{CNT}}{S_{CNT}} + \sqrt{1 + \frac{L_{CNT}^2}{S_{CNT}^2}} \right) - \sqrt{1 + \frac{S_{CNT}^2}{L_{CNT}^2} + \frac{S_{CNT}}{L_{CNT}}} \right]. \tag{8.9}$$

The kinetic inductance (quantum inductance) is given by:

$$L_{kin} = \frac{h}{4n_{CNT}e^2 v_f}. \tag{8.10}$$

Variables used in CNT-TSV modeling are summarized in Table 8.1; typical values of parasitic components for Cu-TSV and CNT-TSV are shown in Table 8.2.

Table 8.1 Variables used in CNT-TSV modeling

Parameter	Symbol
The equivalent resistance	R_{eqCNT}
Scattering effect resistance	R_{scat}
Spin and sub-lattice degeneracy of electrons	R_{deg}
Number of CNT in bundle	n_{CNT}
Height of CNT	L_{CNT}
Diameter of a single SW-CNT	D_{CNT}
Mean free path of electrons in CNT	λ
Fermi velocity of CNT	v_f
The equivalent capacitance	C_{eqCNT}
Quantum capacitance	C_{qnt}
Electrostatic capacitance	$C_{electro}$
The equivalent inductance	L_{eqCNT}
Magnetic inductance	L_{mag}
Kinetic inductance	L_{kin}

Table 8.2 Typical values of parasitic components for Cu-TSV and CNT-TSV per 100 μm

Technology	Cu-TSV	CNT-TSV
Resistance (mΩ)	10–60	300
Inductance (pH)	5–10	0.005–0.008
Capacitance (fF)	3–10	0.1–0.3

S parameter simulations show that S_{21}(CNT_TSV) > S_{21}(Cu_TSV), S_{11}(CNT_TSV) < S_{11}(Cu_TSV) as depicted in Fig. 8.7.

8.3 Adiabatic TSV

Energy recovery logic is a design technique for minimizing the energy dissipation. Its major limitation is the requirement for passive components, which cannot be efficiently integrated into current generation integrated circuits (ICs). TSV-based 3D heterogeneous integration may enable efficient integration of these passive elements, which were not practically feasible in the past due to technology limitations [9–14]. TSVs can provide better inductors needed for adiabatic circuits.

8.3.1 Adiabatic Theory

The word adiabatic comes from a Greek word that is used to describe thermodynamic processes that exchange no energy with the environment and therefore, no energy loss in the form of dissipated heat. In real-life computing, such an ideal process cannot be achieved because of the presence of dissipative elements like resistances in a circuit. However, one can achieve very low energy dissipation by slowing down the speed of operation and only switching transistors under certain

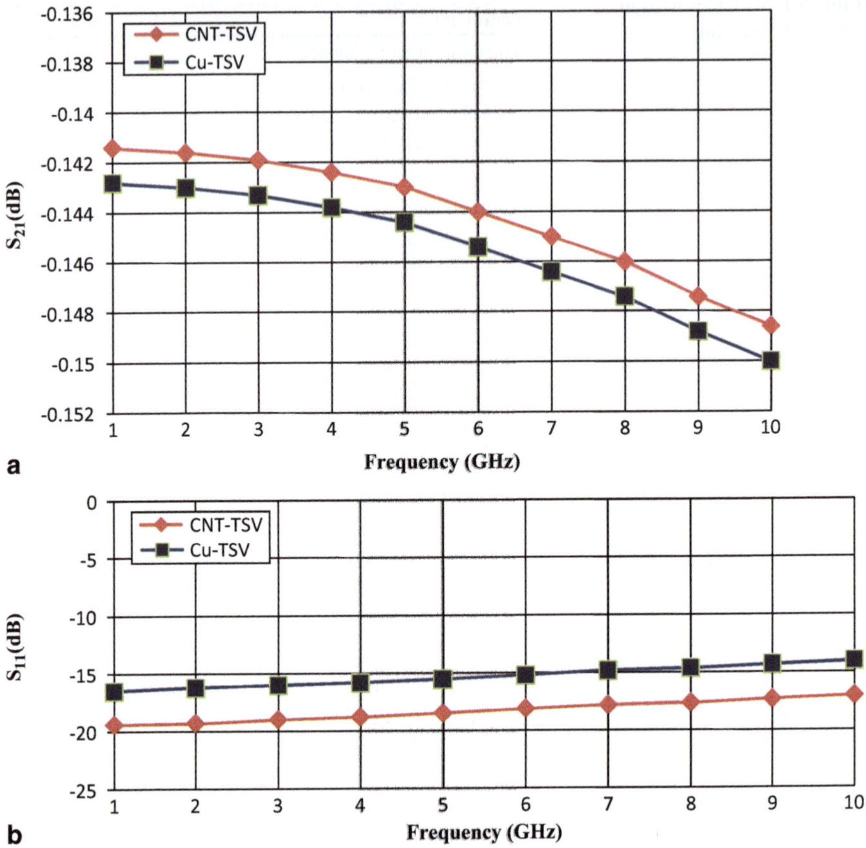

Fig. 8.7 **a** S_{21} parameter of both Cu-TSV and SW-CNT-TSV versus frequency, **b** S_{11} parameter of both Cu-TSV and SW-CNT-TSV versus frequency

conditions. The signal energies stored in the circuit capacitances are recycled instead of being dissipated as heat. The adiabatic logic is also known as energy recovery complementary metal oxide semiconductor (CMOS). To reduce the energy dissipation in conventional CMOS, either the supply voltage V_{DD} has to be reduced or the load capacitance C. In contrast to the conventional CMOS circuit, the supply voltage in adiabatic case is not constant, but time variable with a slow rising time ($T \gg RC$). The derivation is as follows [15–18]:

$$E = PT \tag{8.11}$$

$$P = I^2 R \tag{8.12}$$

$$Q = IT \tag{8.13}$$

$$Q = CV \tag{8.14}$$

From (8.11), (8.12), (8.13) into (8.14) results in:

$$E_{adiabatic} = \left(\frac{RC}{T}\right) CV_{DD}^2. \tag{8.15}$$

Noting that:

$$E_{conventional} = (\tfrac{1}{2})CV_{DD}^2, \tag{8.16}$$

where the various terms of Eq. (8.15) are described as follows:

E Energy dissipated during charging
Q Charge being transferred to the load
C Value of the load capacitance
R Resistance of the MOS switch turned on
V Final value of the voltage at the load
T Time spent for charging

Now, a number of observations can be made based on Eq. (8.15) as follows:

1. The dissipated energy is smaller than for the conventional case, if the charging time *T* is larger than *2RC*. That is, the dissipated energy can be made arbitrarily small by increasing the charging time,
2. Also, the dissipated energy is proportional to *R*, as opposed to the conventional case, where the dissipation depends on the capacitance and the voltage swing. Thus, reducing the on-resistance of the P-type metal oxide semiconductor (PMOS) network will reduce the energy dissipation.

Phases in an adiabatic power supply are shown in Fig. 8.8, where charges move efficiently from power supply to the load capacitance by using slow and constant current charging. Reversing the current source will cause the energy to flow from the load capacitance back into the power supply. The power supply is designed to retrieve the energy fed back to it. Adiabatic switching circuits require nonconstant, nonstandard power supply with time-varying voltage. This supply is called pulse power supplies.

8.3.2 Adiabatic Architecture for 3D Integration

The adiabatic architecture is introduced, where it should contain CMOS to adiabatic converter, adiabatic to CMOS converter, pulse voltage generator, and function logic. The differences between conventional TSV and adiabatic TSV are shown in Fig. 8.9.

8.3.2.1 Pulse Voltage Generator

The circuit used as a pulse voltage generator is shown in Fig. 8.10. In an adiabatic circuit, the power supply also acts as a clock. Reduction of the power dissipation occurs at the cost of slower speed of operation [15]. The inductor is implemented using TSV technology.

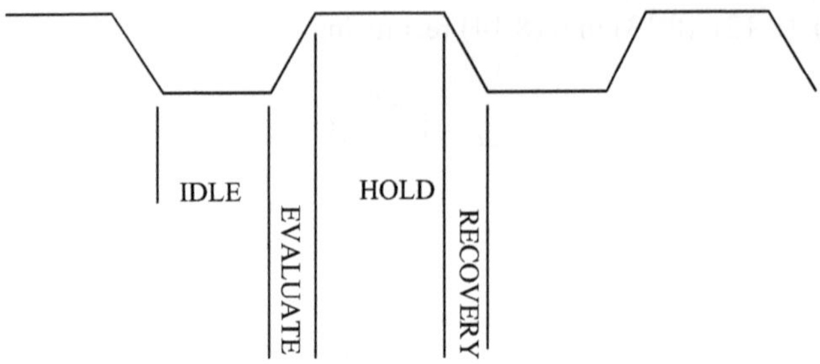

Fig. 8.8 Phases in an adiabatic power supply, where charges move efficiently from power supply to the load capacitance by using slow and constant current charging

Fig. 8.9 **a** Conventional TSV, **b** adiabatic TSV

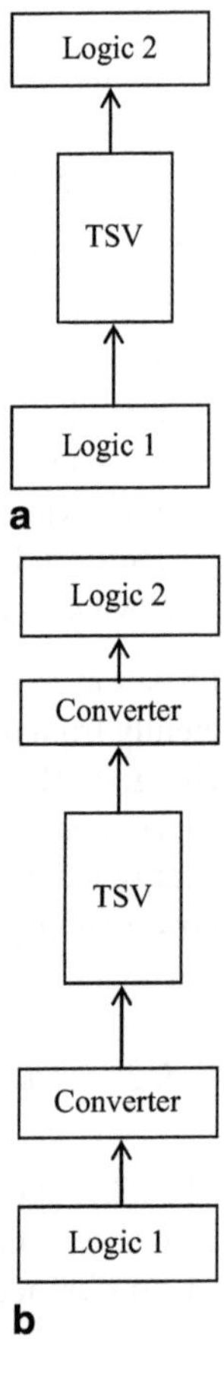

8.3.2.2 CMOS to Adiabatic Converter

To convert a conventional CMOS logic gate into an adiabatic gate, the pull-up and the pull-down networks must be replaced with complementary transmission-gate

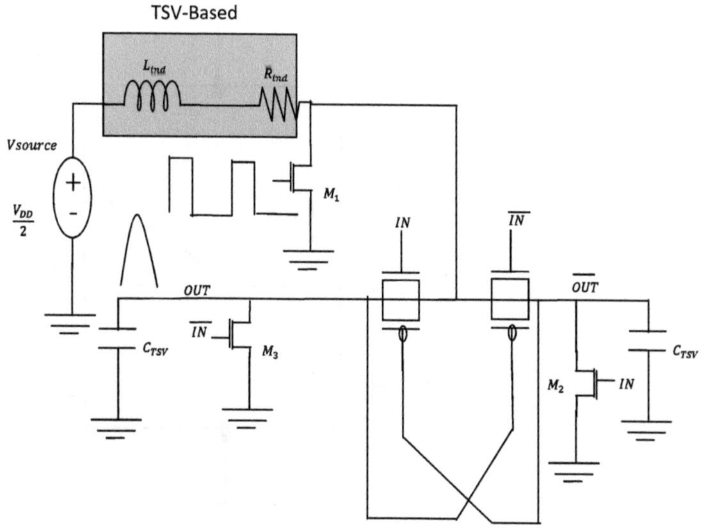

Fig. 8.10 Pulse voltage generator [15]

Fig. 8.11 CMOS to adiabatic converter [15]

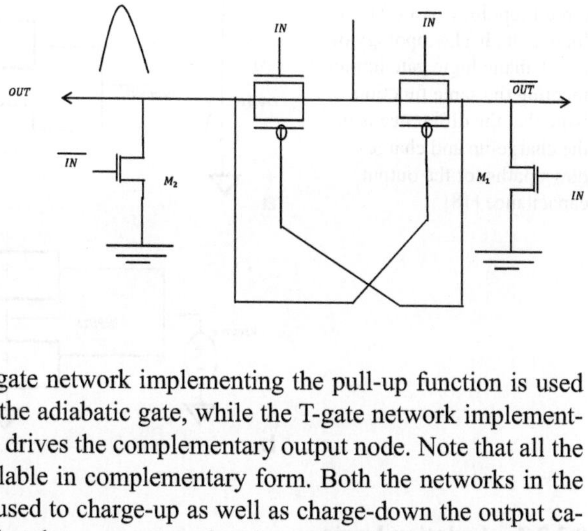

(*T-gate*) networks. The T-gate network implementing the pull-up function is used to drive the true output of the adiabatic gate, while the T-gate network implementing the pull-down function drives the complementary output node. Note that all the inputs should also be available in complementary form. Both the networks in the adiabatic logic circuit are used to charge-up as well as charge-down the output capacitance, which ensures that the energy stored at the output node can be retrieved by the power supply, at the end of each cycle. To allow adiabatic operation, the DC voltage source of the original circuit must be replaced by a pulsed-power supply with the ramped voltage output (Fig. 8.11).

8.3.2.3 Adiabatic to CMOS Converter

To convert an adiabatic CMOS logic gate into a conventional gate, we use the circuit shown in Fig. 8.12 [15].

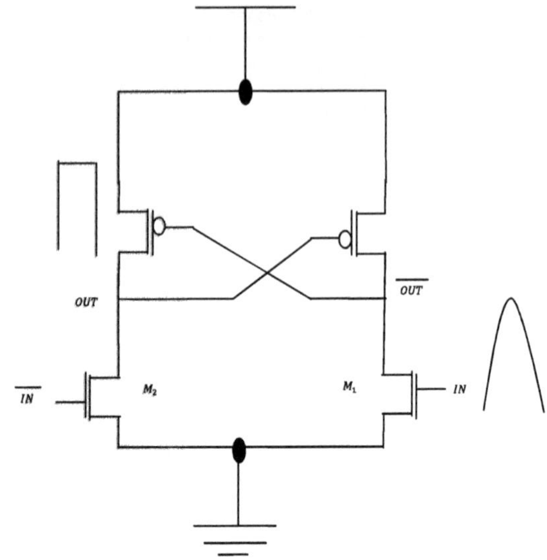

Fig. 8.12 Adiabatic to CMOS converter [15]

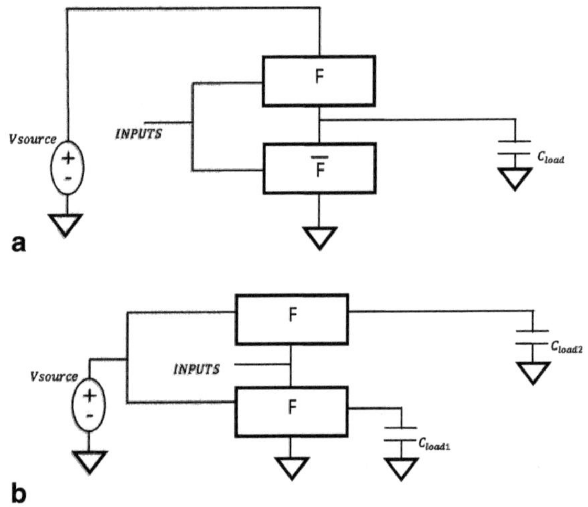

Fig. 8.13 **a** The general circuit topology of a CMOS logic gate. **b** The topology of an adiabatic logic gate implementing the same function. Note that the difference is in the charge-up and charge-down paths for the output capacitance [18]

8.3.2.4 Function Logic

To implement a logic function in adiabatic logic: replace each of the PMOS and NMOS devices in the pull-up and pull-down networks with T-gates. Use the pull-up network to drive the true output and the pull-down network to drive the complementary output. Both networks are used to charge and discharge the load capacitance. The general circuit topology of a CMOS logic gate and the topology of an adiabatic logic gate implementing the same function are shown in Fig. 8.13. Note that the difference is in the charge-up and charge-down paths for the output capacitance [18].

Table 8.3 Variables comparison of power consumption between conventional CMOS and adiabatic logic for NAND, NOR, and ADDER implementation

Circuit	Conventional CMOS (nW)	TSV-based adiabatic (nW)	Improvement (%)
NAND	1338	559	58
NOR	1157	355	70
ADDER	475×10^4	185×10^4	61

8.4 Simulations and Comparison with CMOS

Comparison between conventional CMOS and TSV-based adiabatic logic regarding the power consumption is presented in Table 8.3, where a significant improvement in the power in noticed. Note that the inductor in the pulse voltage generator is implemented using TSV technology.

8.5 Summary

In this chapter, performance comparison between air-gap-based coaxial TSV and conventional circular TSV were presented. Air-gap TSVs reduce the energy loss compared to the conventional circular TSV or conventional coaxial TSV. However, the air-gap TSV maybe challenge in fabrication. Moreover, SW-CNT bundles as a prospective filler material for TSV were investigated compared to conventional filler materials like Cu, W, and polysilicon. CNT-filled TSVs have superior performance compared to Cu-filled TSVs, resulting in reduction in transmission loss at high frequency of operation. Moreover, energy recovery logic is a design technique for minimizing the energy dissipation. Its major limitation is the requirement for passive components, which cannot be efficiently integrated into current generation ICs. TSV-based 3D heterogeneous integration may enable efficient integration of these passive elements, which were not practically feasible in the past due to technology limitations. In this work, TSV-based adiabatic logic based on the adiabatic switching principle was presented and analyzed.

References

1. L. Yu, H. Yang, J. Zhang, W. Wang, Performance evaluation of air-gap-based coaxial RF TSV for 3D NoC. IEEE/IFIP 19th international conference on VLSI and system-on-chip, 2011
2. L. Yu, H. Yang, T.T. Jing, M. Xu, R. Geer, W. Wang, Electrical characterization of RF TSV for 3D multi-core and heterogeneous ICs. IEEE, 2010
3. A. Gupta, B.C. Kim, S. Kannan, S.S. Evana, L. Li, Analysis of CNT based 3D TSV for emerging RF applications. Electronic components and technology conference, 2011
4. B. Kim, S. Kannan, A. Gupta, S. Noh, L. Li, Characterization of high performance CNT-based TSV for radar applications. IEEE 13th electronics packaging technology conference, 2011

5. S. Kannan, B. Kim, A. Gupta, S. Noh, L. Li, Characterization of high performance CNT-based TSV for high-frequency RF applications. Adv. Mater. Res. **1**(1) (2012)
6. T. Wang, K. Jeppson, N. Olofsson, E. Campbell, J. Liu, Through silicon vias filled with planarized carbon nanotube bundles. IOPscience, Nanotech. **20**(48), (2009)
7. A. Gupta, S. Kannan, B. Kim, F. Mohammed, B. Ahn, Development of novel carbon nanotube TSV technology. IEEE electronic components and technology conference, 2010
8. S. Hassan, M. Asghar, Limitation of silicon based computation and future prospects. Second international conference on communication software and networks, 2010
9. N. Arora, B.P. Singh, T. Sharma, K.G. Sharma, Adiabatic and standard CMOS interfaces at 90 nm technology. WSEAS Trans. Circuits Syst. **9**(3) (2010)
10. J.C. Garcia, J.A. Montiel, S. Nooshabadi, A CMOS adiabatic inverter operating with a single clock power supply to reduce non-adiabatic loss. IEEE, 2008
11. P. Asimakopoulos, G. Van der Plas, A. Yakovlev, P. Marchal, Evaluation of energy-recovering interconnects for low-power 3D stacked ICs. IEEE, 2009
12. P.D. Khandekar, S. Subbaraman, A.V. Chitre, Implementation and analysis of quasi-adiabatic inverters. IMECS, 2010
13. H. Mahmoodi, V. Tirumalashetty, M. Cooke, K. Roy, Ultra low-power clocking scheme using energy recovery and clock gating. IEEE Trans. Very Large Scale Integr. (VLSI) Syst. **17**(1), 33–44 (2009)
14. P. Sreenivasulu, M.V. Narasimha, G.V.K. Varaprasad, Design and analysis of low power full adder using adiabatic technique. Int. J. Emerg. Technol. Adv. Eng. **2**(6) (2012)
15. S. Mandavilli, P. Paramahans, An efficient adiabatic circuit design approach for low power applications. Int. J. Recent Trends Eng. **2**(1) (2009)
16. N. Anuar, Y. Takahashi, T. Sekine, Adiabatic logic versus CMOS for low power applications. IEEE, 2010
17. N. Anuar, Y. Takahashi, T. Sekine, Two phase clocked adiabatic static logic circuit: a proposal for digital low power applications. Proc. IEICE General Conference, March 2009, p. 102
18. R. Kumbhare, J. Kanungo, A.K. Saxena, S. Dasgupta, Design of an ultra low power clock gating D flip-flop using quasi-static energy recovery logic. Microelectron. Solid State Electron. **1**(1), 9–14 (2012)

Chapter 9
TSV Fabrication

3D integration is a promising and attractive solution for interconnect bottleneck problem, transistor scaling physical limitations, and impractical small-scale lithography. 3D integration extends Moore's law in the third dimension, offering heterogeneous integration, higher density, lower power consumption, and faster performance. However, in order to fabricate 3D integrated circuits (ICs), new capabilities are needed: process technology, physical modeling, physical design tools, 3D architectures, design methods, and tools. The goal of this chapter is to cover the manufacturability of through silicon via (TSV)-based 3D-ICs, i.e., process technology and fabrication capability.

9.1 TSV Fabrication Process Overview

There exist five key components for 3D integration technology.

9.1.1 Stacking

The main types of stacking are wafer-to-wafer (W2W), die-to-wafer (D2W), and die-to-die (D2D) stacking (Table 9.1); each type has its own advantages [1]. Selection of the appropriate type depends on many factors such as die size, delay, yield, and fabrication constraints.

1. For *D2D* stacking, the dies are cut from different similar/dissimilar wafers. Normally, the dies are tested individually before stacking together. The dies can be stacked using wire bonding, solder bumps, or through silicon via (TSVs). The dies can be mixed; for example, a faster memory die can be matched to the faster logic die. D2D stacking can be a good option for heterogeneous integration where the wafer sizes may be significantly different. But alignment of the dies can be a major problem in this approach. This approach is not cost-effective compared to

K. Salah et al., *Arbitrary Modeling of TSVs for 3D Integrated Circuits*, Analog Circuits and Signal Processing, DOI 10.1007/978-3-319-07611-9_9,

Table 9.1 Stacking technologies

	Wafer-to-wafer	**Die-to-wafer**	**Die-to-die**
Figure	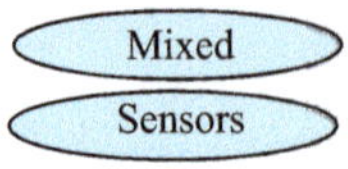	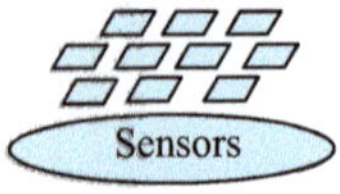	
Advantages	1-Low cost 2-High production throughput	1-easy to integrate heterogeneous technologies. 2-high yield due to flexible, use of KGD	1- High yield due to flexible, use of KGD.
Disadvantages	1-Difficult to integrate heterogeneous technologies. 2-low yield as there is no control over attaching a bad to a good die.	1-Handling and bonding 2-ESD (Electrostatic discharge): where the die pick-and-place process is vulnerable to a voltage potential difference between itself and the external environments.	1-Not feasible for mass production 2-Handling and bonding 3-ESD

the W2W stacking. However, known good dies (KGD) can be stacked together using this approach.

2. In the case of *D2W* stacking, it is almost similar to D2D stacking except that dies are stacked to a substrate. D2W stacking has the advantage of flexible use of KGD. KGD are stacked over a substrate wafer; for example, using die-cavity technology. It is suitable for common die sizes.
3. For W2W stacking, the wafers are stacked together in W2W stacking. It has the advantage of low cost. Some good dies may be forced to bond to bad dies. The yield can be a big issue for W2W stacking. This approach is suitable for common wafer size.

9.1.2 Bonding

3D stacking can be carried out using three main bonding techniques; face-to-face (F2F), face-to-back (F2B), and back-to-back (B2B; Table 9.2), where "face" refers to the active front side of the die, and "back" refers to the substrate backside [2–3].

Table 9.2 Bonding technologies

	Face-to-face	Face-to-back	Back-to-back
Figure	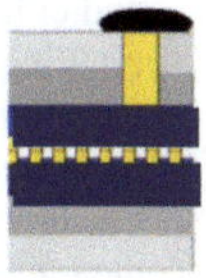		
Advantages	1-Few TSVs are required.	1-Allows an arbitrary number of dies to be stacked.	1-Bonding process is easier.
Disadvantages	1-Allows only two dies to be stacked.	1-More TSVs are required.	1-More TSVs are required. 2-allows only two dies to be stacked.

1. For *F2F,* although F2F bonding allows only two dies to be bonded, it can provide higher TSV density due to similarity of synthesizing on die interconnects. F2F bonding technique can be very useful for W2W bonding in order to avoid any through hole through the delicate wafer.
2. For *F2B,* the active layer of one die faces the back or substrate side of the other die in F2B bonding. However, any number of dies can be stacked using the F2B approach but TSV density is less than its F2F counterpart as TSV must be etched through the back of the die and the etching process has low resolution compared to the F2F bonding approach. The F2B bonding technique is suitable for D2D or D2W bonding.
3. For *B2B,* the substrate sides of both the dies are facing each other in B2B bonding. It is rarely used because of the long vertical interconnect between a pair of neighboring dies.

9.1.3 TSV Formation

This can be categorized into via-first and via-last approach (Table 9.3) [4]. If TSVs are introduced quite early in the process even before devices are built, it is a called via-first process. If TSVs are introduced after bonding and thinning then it is called a via-last process.

Table 9.3 TSV formation schemes

	Via-first	Via-last
Fabrication	Fig. 9.1	Fig. 9.2
Description	1. "Via first" formation is done either before (FEOL: front end of line) or after FEOL CMOS devices processing. Mainly, "via first" means fabrication of TSVs before stacking of devices 2. Via-first TSVs occupy the device (= placement) layer	1. "Via-last" approach occurs after wafer fabrication (FEOL + BEOL) and either before or after wafer bonding. Mainly, "via last" means fabrication of TSVs after stacking of devices 2. Via-last TSVs occupy both the device and metal (= routing) layers
Advantages	1. TSV density is high 2. BEOL routability is good	1. Cost is low 2. Both wafer and die stacking 3. Lower interconnection density
Disadvantages	1. Cost is high 2. Wafer stacking only 3. Higher interconnection density	1. TSV density is low 2. BEOL routability is acceptable 3. Process complexity is high

CMOS complementary metal oxide semiconductor

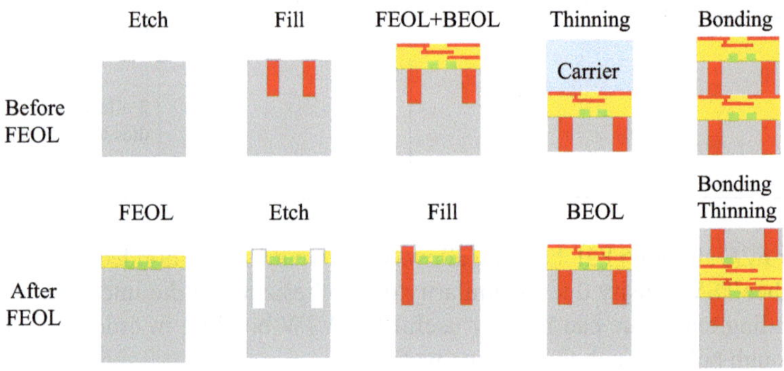

Fig. 9.1 Via-first process

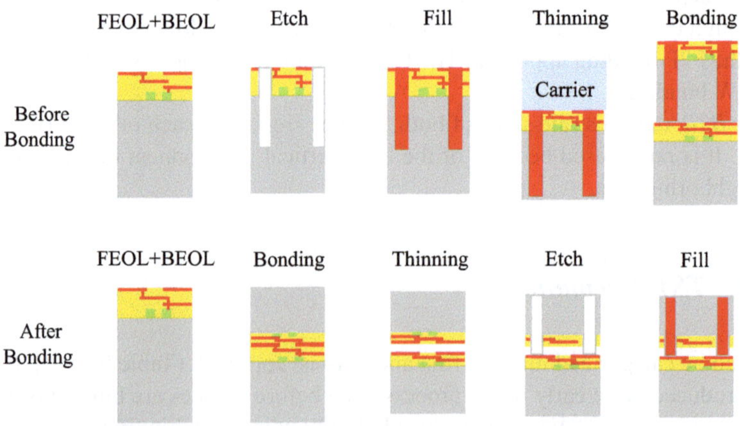

Fig. 9.2 Via-last process

1. In the *via-first* approach, the TSV is etched and filled first and then the transistor is fabricated at high temperature. In the next process phase, metal contacts are made, the back is thinned using a carrier, and the carrier is debonded at the end. Devices and TSVs are fabricated before deposition of metal layers and bonding of dies in the via-first approach. In via-first technology, the TSVs are surrounded laterally by other TSVs and vertically by wires.
2. In the *via-last* approach, front end of line (FEOL) takes place at high temperature, and then back end of line (BEOL) takes place at relatively low temperature. Then thinning takes place, after this step TSV etching and filling are done, and finally debonding of the carrier takes place.

9.1.4 Alignment

The alignment accuracy depends on stacking and bonding techniques mentioned above [5–8]. The alignment tolerances can have a direct impact on the density of connections achievable in the 3D stack, and thus the overall performance.

Tolerances can vary widely depending on the tooling and process flow selected, so it is important to be aware of process capabilities. For example, D2D assembly processes can have alignment tolerances varying from the 1-μm range to the 20-μm range, depending on the speed of assembly required.

9.1.5 Thinning

The substrate can be in either silicon on insulator (SOI) or bulk silicon; therefore, the thinning should be done for both. The benefit of the substrate thinning is to reduce the TSV resistance, but the dimension of the TSVs is not expected to scale at the same rate as feature size because alignment tolerance during bonding poses limitation on the scaling of the vias [9].

Wafer-thinning processes which use handle wafers and lamination can add distortion to the thinned silicon layer. This distortion can be caused by both differences in coefficients of thermal expansion between materials and by the use of polymer lamination materials with low elastic modulus.

Wafer thinning is a necessary component of 3D integration as it allows the interlayer distance to be reduced, therefore allowing a high density of vertical interconnects. The big challenge in wafer thinning is that the wafer must be thinned to ~5–10 % of its original thickness. This thinning is especially challenging since there is no natural etch stop.

An example of TSV-based 3D-ICs fabrication process is "SOI/W2W/F2F/via-last process," where the first part (SOI) determines the substrate type and the second (W2W) and third part (F2F) determine the stacking and bonding technologies, respectively. Finally, the TSV formation process is determined by the fourth part (via last).

There are many Fabs for TSV-based 3D integration ICs like, ALLVIA, Suss [10], and Austria microsystems [11].

9.2 The Fabrication Process Flow

The fabrication process flow of the 3D TSV is as follows:

9.2.1 TSV Etching

A 6-inch silicon wafer (bulk) is used for the TSV fabrication with resistivity>104 Ω-cm to reduce the substrate loss. The first process in TSV fabrication is TSV etching. TSV structures can be fabricated by a number of methods of silicon removal, such as wet processing using a chemical etch, or by dry etching, such as laser drilling, reactive ion etch (RIE), or deep reactive ion etching (DRIE). Combinations of these etch processes are also possible. Etching often begins with photolithography and the use of a photoresist or a "hard mask," such as a patterned oxide or nitride layer to define the features to be etched.

9.2.2 Oxide Deposition

Oxide deposition is needed to coat the sidewalls of the TSV to ban current leakage to substrate, i.e., for electrical isolation [12]. TSVs may be fabricated with or without dielectric insulation from the silicon. For simple structures with a desired short to the base silicon, no dielectric is required. For application requiring electrical isolation from the silicon, a polymer dielectric or oxide or nitride dielectric may be used.

9.2.3 TSV Filling

Deposition of a conductor begins with deposition of a thin conductive liner and seed layer.

Examples of TSV conductor fill materials include polysilicon, copper, tungsten. Poly Si is compatible with FEOL and Cu is compatible with BEOL. Techniques for TSV filling include electroplating or thermal paste [12–13].

Table 9.4 TSV formation equipment: micromachining technology

Equipment	Process	Figure
Spinner	Lithography: photoresist spin coating	Fig. 9.3
Aligner	Lithography: to align mask to wafer	Fig. 9.4
Deep reaction ion etching (DRIE)	TSV etching	Fig. 9.5
Plasma enhanced chemical vapor deposition (PECVD)	Oxide deposition	Fig. 9.6
Electroplater	TSV filling	Fig. 9.7
Chemical mechanical polisher (CMP)	Substrate thinning	Fig. 9.8

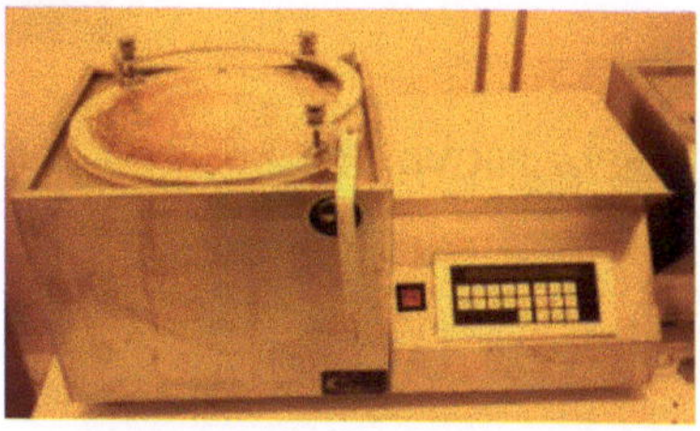

Fig. 9.3 Spinner [15]

Fig. 9.4 Aligner [15]

9.2.4 Substrate Thinning

In order to thin the wafer, the device wafer (*Si* interposer) is temporally bonded onto a carrier wafer [13–15]. The benefit of the substrate thinning is to reduce the TSV resistance.

The corresponding equipment used in TSV fabrication are summarized in Table 9.4 [13]. Applications for TSV-based 3D include image sensors, flash memories, DRAM, processors, field-programmable gate arrays (FPGAs), and power amplifiers. Table 9.5 provides examples of company announcements. The timing for mass production depends on how the TSV compares in terms of cost with existing technologies.

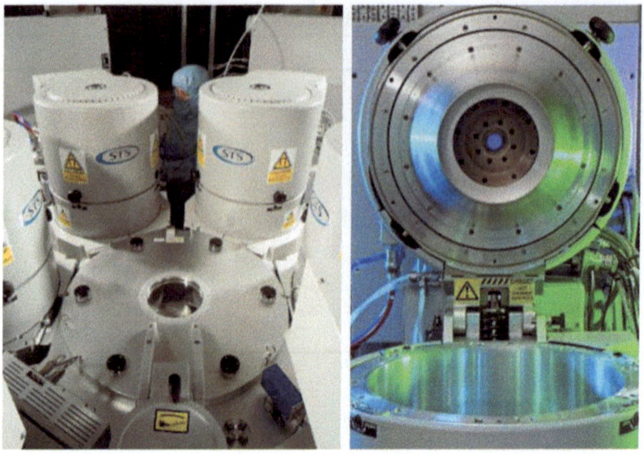

Fig. 9.5 DRIE [15]

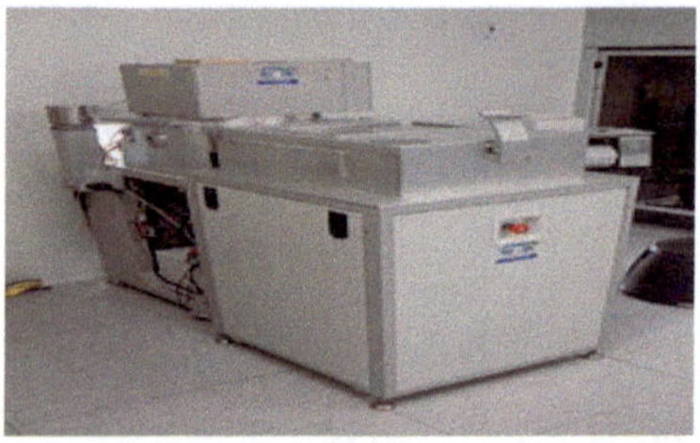

Fig. 9.6 PECVD [15]

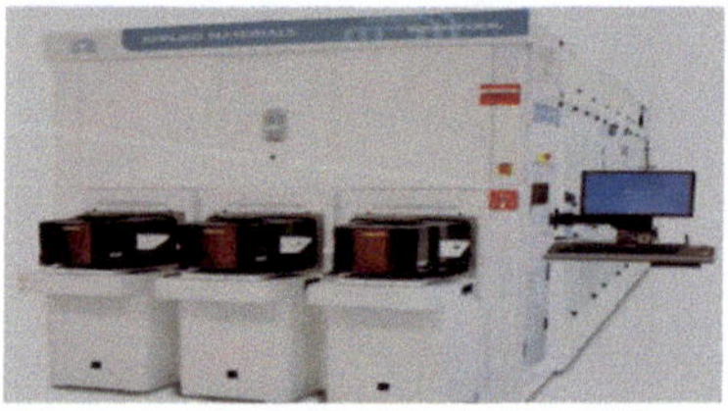

Fig. 9.7 Electroplater [15]

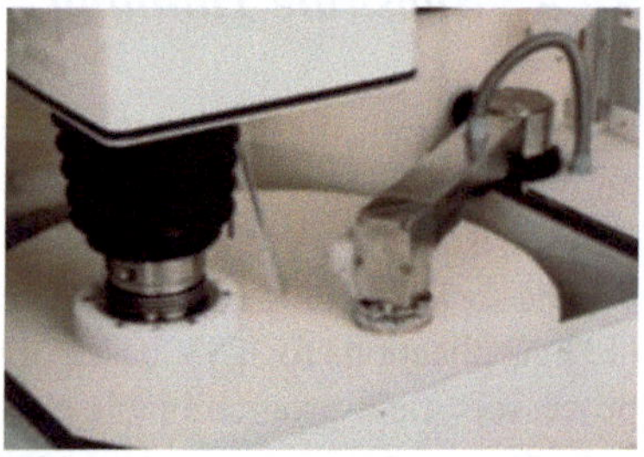

Fig. 9.8 CMP [15]

Table 9.5 3D TSV products in the near future

Company	Memory and logic	RAM	Flash	Image sensor	Power amplifier	FPGA
IBM	✓				✓	
Intel	✓					
Xilinx						✓
Toshiba				✓		
Samsung		✓	✓			
Tezzaron		✓				
Sony	✓					
Sanyo				✓		
Fujikura				✓		
Micron		✓	✓	✓		
ZyCube				✓		
Tessera				✓		

9.3 Summary

This chapter presented the TSV fabrication process. It covered the manufacturability of TSV-based 3D-ICs, i.e., process technology and fabrication capability. Moreover, it discussed the five key components for 3D integration technology: staking, bonding, TSV formation, alignment, and thinning.

References

1. H.U. Knickerbocker, P.S. Andry, B. Dang, R.R. Horton, C.S. Patel, R.J. Polastre, K. Sakuma, E.S. Sprogis, C.K. Tsang, B.C. Webb, S.L. Wright, 3D silicon integration. Electronic components and technology conference, 2008
2. Y. Xie, Processor architecture design using 3D integration technology. 23rd international conference on VLSI design, 2010
3. E.J. Marinissen, Y. Zorian, Testing 3D chips containing through-silicon vias. International test conference, 2009
4. T.M. Bauer, S.L. Shinde, J.E. Massad, D.L. Hetherington, Front end of line integration of high density, electrically isolated, metallized through silicon vias. Electronic components and technology conference, 2009
5. D.Y. Chen, W.C. Chiou, M.F. Chen, T.D. Wang, K.M. Ching, Enabling 3D-IC foundry technologies for 28 nm node and beyond: through-silicon-via integration with high throughput die-to-wafer stacking. IEEE, 2009
6. A. Klumpp, P. Ramm, R. Wieland, 3D-integration of silicon devices: a key technology for sophisticated products. Date, 2010
7. C. Laviron, B. Dunne, V. Lapras, P. Galbiati, D. Henry, F. Toia, S. Moreau, R. Anciant, C.B. Manquat, N. Sillon, Via first approach optimisation for through silicon via applications. Electronic components and technology conference, 2009
8. R. Weerasekera, M. Grange, D. Pamunuwa, On signalling over through-silicon via (TSV) interconnects in 3-D integrated circuits. Date, 2010

9. M. Koyanagi, T. Fukushima, T. Tanaka, Three-dimensional integration technology and integrated systems. IEEE, 2009
10. http://www.suss.com/markets/3d-integration/tsv-manufacturing.html. Accessed 2012
11. http://www.austriamicrosystems.com/. Accessed 2012
12. Y. Civale, D.S. Tezcan, H.G.G. Philipsen, P. Jaenen, R. Agarwal, F. Duval, P. Soussan, Y. Travaly, E. Beyne, Die stacking using 3D-wafer level packaging copper/polymer through-si via technology and Cu/Sn interconnect bumping. IEEE, 2009
13. M.J. Wolf, P. Ramm, A. Klumpp, H. Reichl, Technologies for 3D wafer level heterogeneous. DTIP, 2008
14. S.Q. Gu, U. Ray, Y. Li, A. Chandrasekaran, B. Henderson, M. Nowak, 3D TSV integration technology challenges for high volume production from fabless supply chain aspect. IEEE, 2010
15. C. Huyghebaert, J.V. Olmen, Y. Civale, A. Phommahaxay, A. Jourdain, S. Sood, S. Farrens, P. Soussan, Cu to Cu interconnect using 3D-TSV and Wafer to Wafer thermo compression bonding. IEEE, 2010

Chapter 10
Conclusions

Scaling of complementary metal-oxide semiconductor (CMOS) technology is saturating soon. This will definitively cause the seizure of development in computing if no alternatives are found. These alternatives should have the capacity to overcome all limitations of current techniques. These alternatives are silicon on insulator (SOI), FinFET, twin-well, high-*K*, metal gate, strained-Si, 3D, network on chip (NoC), optical, wireless, molecular computer, biological computer, and quantum computer. All these new technologies are under development and may be available for commercial use in the near future. Three-dimensional integration technology using through silicon via (TSV) is one of the possible future integrated circuit (IC) design technologies. It can offer many advantages over the existing 2D integration technology. However, many challenges (modeling, thermal, computer-aided design (CAD) tools, yield and test, technological, design methodologies, circuit architectures) should be overcome before TSV-based 3D IC production becomes possible.

In this book, a TSV modeling methodology was introduced. The methodology fits a physics-based lumped element model which includes skin effects, nonlinear TSV capacitance, coupling parasitics, adjacent body contact effects, as well as the parasitic elements due to the finite substrate resistivity to the complete frequency-dependent TSV empirical data obtained from a field solver. Both 3D full-wave (high frequency structure simulator, HFSS) and 3D quasi-static (Q3D) field solvers were considered. Comparison between those two field solvers from few Hz to 100 GHz showed that the difference in accuracy is $<5\,\%$; hence, the 3D Quasi-static (Q3D) field solver was used. Dimensional analysis was used to develop closed-form formulae for the lumped elements. The closed-form formulae achieved a very good agreement with the field solver *S* parameter with error $<5\,\%$. The model was used to study the impact of a typical TSV on a 3D circuit performance. This work showed that the self-resistance and inductance of the TSV have negligible effect on high-impedance device characteristics; hence in such a case, the TSV can be considered a capacitance-dominated structure. Also, this work showed that the TSV self-inductance and resistance have a significant effect on low-impedance device characteristics. This book also shows that the capacitance of the TSV is highly dependent on the body contact placement and count. Also, it is shown that the inductive and resistive coupling between TSVs can be neglected and the coupling network between

K. Salah et al., *Arbitrary Modeling of TSVs for 3D Integrated Circuits*,
Analog Circuits and Signal Processing, DOI 10.1007/978-3-319-07611-9_10,

TSVs can be reduced to only capacitors. Moreover, the TSV nonlinear elements were represented with multiple linear elements. Furthermore, the TSV model was verified using device simulations with error <2 %.

Moreover, this book presented a detailed TSV structure and the challenges that should be overcome before the production of TSV-based 3D ICs becomes possible, e.g., electrical modeling challenges, thermal and power challenges, technological challenges, design methodology challenges, and CAD tool development challenges. In this book, a complete set of self-consistent equations including self and coupling terms for capacitance of general multi-TSV structures was introduced. The error when using the closed-form expressions as compared to a field solver is <7 %. TSV parasitic capacitance is less than other conventional input/output (IO) structures' capacitance; therefore, it is suitable for low-power applications.

Moreover, an architecture based on TSV technology for a spiral inductor was demonstrated and characterized. A closed-form expression for the inductance was obtained and an analytical equation to calculate the maximum quality factor was derived. Simulation results indicate that our formulas are accurate. Moreover, simulations showed that our model presents better quality factor than planar spiral inductor (120 % improvement) and 3D via-based spiral inductor (76 % improvement) for identical inductance. Our inductors have *Q*s in the 8–21 range. Moreover, the self-resonance frequency of the TSV-based 3D spiral inductor is 38 % higher than the conventional 2D spiral inductor and 3 % higher than the 3D via-based inductor. It also provides lower area for the same inductance, where the inductor occupies only 15 % of the area of the conventional 2D planar spiral inductor and 60 % of the area of 3D via-based spiral inductor for the same inductance and higher quality factor. These good characteristics of the inductor indicate that TSV/3D integration is a promising solution for radio frequency system on chip (RF-SoC).

Also, in this book, and for the first time, a novel inductive coupling interface that uses the magnetic near field induced by TSV-based spiral inductor was demonstrated. The feasibility of using TSV for wireless near-field communication was shown. A TSV-based near-field inductive-coupling system offers high-quality factor and good coupling coefficient. Simulations predict that the transceiver fed with a 1-Gbps input data stream can maintain very small jitter at the output. The wireless communication link provides small area $(30\ \mu m)^2$ for typical design values. Therefore, the communication system appears to be a promising technology for wireless communication. To our knowledge, this is the first work on this topic. Moreover, a structure based on TSV technology for a bandpass filter was demonstrated and characterized.

In this book, the effects of substrate doping density on the electrical performance of TSV were investigated. The previously introduced lumped circuit model for TSV structure was used for a lightly doped silicon structure. A new lumped circuit model for a heavily doped silicon substrate was introduced and its physical understanding was explained. Both circuit models for the lightly doped and heavily doped cases were validated using full-wave simulations up to 10 GHz. An equivalent lumped model for more than two stratified layers can be obtained by cascading the appropriate model according to the substrate type. Moreover, parasitic coupling capacitive between TSVs and surrounding wires was analyzed and studied, where our model considers coupling with surrounding wires in vertical directions. The simulation results for the electrical characteristics of the coupling between TSVs and metal lines in 3D-ICs showed that the coupling is not negligible when TSV is relatively short compared to the TSV width, where the aspect ratio is <5. Therefore, TSV-to-wire capacitance needs to be considered for the computation of TSV capacitance. But, if the aspect ratio is larger than 5, the effect of metal wires is not considered. Moreover, the effect of metal lines on TSV–TSV coupling can be neglected if the pitch is less than three times the TSV diameter. This analysis is very helpful for timing analysis in 3D-ICs. Moreover, in this chapter, noise coupling between TSVs and CMOS was investigated. Several isolation techniques were also introduced.

Performance comparison between air-gap-based coaxial TSV and conventional circular TSV were presented. The air-gap TSVs reduce the energy loss compared to the conventional circular TSV or conventional coaxial TSV. But, air-gap TSV may be a challenge in fabrication.

Single-walled carbon nanotube (SW-CNT) bundles as a prospective filler material for TSV were investigated compared to conventional filler materials like Cu, W, and polysilicon. CNT-filled TSVs have superior performance compared to Cu-filled TSVs, resulting in reduction in transmission loss at high frequency of operation. Moreover, this book presented the TSV fabrication process.

Index

K. Salah et al., *Arbitrary Modeling of TSVs for 3D Integrated Circuits*,
Analog Circuits and Signal Processing, DOI 10.1007/978-3-319-07611-9,

MIX
Papier aus verantwortungsvollen Quellen
Paper from responsible sources
FSC® C105338

If you have any concerns about our products,
you can contact us on
ProductSafety@springernature.com

In case Publisher is established outside the EU,
the EU authorized representative is:
Springer Nature Customer Service Center GmbH
Europaplatz 3, 69115 Heidelberg, Germany

Printed by Libri Plureos GmbH
in Hamburg, Germany